ADVANCES IN NONLINEAR POLYMERS AND INORGANIC CRYSTALS, LIQUID CRYSTALS, AND LASER MEDIA

Volume 824

Contents

(continued)

Proc· ·ering

Advar ymers
a s,
Liquid Media

Center for Applied Optics Studies/Rose-Hulman Institute of Technology
Center for Electro-Optics/University of Dayton
Center for Optical Data Processing/Carnegie Mellon University
Institute of Optics/University of Rochester
Jet Propulsion Laboratory/California Institute of Technology
National Synchrotron Light Source/Brookhaven National Laboratory
Optical Sciences Center/University of Arizona
Stanford Synchrotron Radiation Laboratory/Stanford University

20–21 August 1987
San Diego, California

Published by
SPIE—The International Society for Optical Engineering
P.O. Box 10, Bellingham, Washington 98227-0010 USA
Telephone 206/676-3290 (Pacific Time) • Telex 46-7053

SPIE (The Society of Photo-Optical Instrumentation Engineers) is a nonprofit society dedicated to advancing engineering
and scientific applications of optical, electro-optical, and optoelectronic instrumentation, systems, and technology.

The papers appearing in this book comprise the proceedings of the meeting mentioned on the cover and title page. They reflect the authors' opinions and are published as presented and without change, in the interests of timely dissemination. Their inclusion in this publication does not necessarily constitute endorsement by the editors or by SPIE.

Please use the following format to cite material from this book:
Author(s), "Title of Paper," *Advances in Nonlinear Polymers and Inorganic Crystals, Liquid Crystals, and Laser Media*, Solomon Musikant, Editor, Proc. SPIE 824, page numbers (1988).

Library of Congress Catalog Card No. 87-62551
ISBN 0-89252-859-1

Printed in the United States of America.

D
621.36
ADV

Proc ering

Volume 824

Advances in Nonlinear Polymers and Inorganic Crystals, Liquid Crystals, and Laser Media

Solomon Musikant
Chair/Editor

Sponsored by
SPIE—The International Society for Optical Engineering

Cooperating Organizations
Applied Optics Laboratory/New Mexico State University
Center for Applied Optics/University of Alabama in Huntsville
Center for Applied Optics Studies/Rose-Hulman Institute of Technology
Center for Electro-Optics/University of Dayton
Center for Optical Data Processing/Carnegie Mellon University
Institute of Optics/University of Rochester
Jet Propulsion Laboratory/California Institute of Technology
National Synchrotron Light Source/Brookhaven National Laboratory
Optical Sciences Center/University of Arizona
Stanford Synchrotron Radiation Laboratory/Stanford University

**20–21 August 1987
San Diego, California**

Published by
SPIE—The International Society for Optical Engineering
P.O. Box 10, Bellingham, Washington 98227-0010 USA
Telephone 206/676-3290 (Pacific Time) • Telex 46-7053

SPIE (The Society of Photo-Optical Instrumentation Engineers) is a nonprofit society dedicated to advancing engineering
and scientific applications of optical, electro-optical, and optoelectronic instrumentation, systems, and technology.

The papers appearing in this book comprise the proceedings of the meeting mentioned on the cover and title page. They reflect the authors' opinions and are published as presented and without change, in the interests of timely dissemination. Their inclusion in this publication does not necessarily constitute endorsement by the editors or by SPIE.

Please use the following format to cite material from this book:
    Author(s), "Title of Paper," *Advances in Nonlinear Polymers and Inorganic Crystals, Liquid Crystals, and Laser Media,* Solomon Musikant, Editor, Proc. SPIE 824, page numbers (1988).

Library of Congress Catalog Card No. 87-62551
ISBN 0-89252-859-1

Printed in the United States of America.

D
621.36
ADV

ADVANCES IN NONLINEAR POLYMERS AND INORGANIC CRYSTALS,
LIQUID CRYSTALS, AND LASER MEDIA

Volume 824

Conference Committee

Chair
Solomon Musikant
General Electric Company

Cochairs
Larry G. DeShazer
Spectra Technology, Inc.

Garo Khanarian
Hoechst-Celanese Corporation

William A. Penn
General Electric Company

Session Chairs
Session 1—Liquid Crystals and Laser Media
William A. Penn, General Electric Company

Session 2—Molecular and Polymeric Optoelectronic Materials I
Anthony F. Garito, University of Pennsylvania

Session 3—Molecular and Polymeric Optoelectronic Materials II
Garo Khanarian, Hoechst-Celanese Corporation

Session 4—Devices and Applications of the New Materials
R. Lytel, Lockheed Missiles & Space Company, Inc.

Conference 824, *Advances in Nonlinear Polymers and Inorganic Crystals, Liquid Crystals, and Laser Media,* was part of a three-conference program on Synthesis and Evaluation of New Materials held at SPIE's 31st Annual International Technical Symposium on Optical & Optoelectronic Applied Science & Engineering. The other conferences were

Conference 822, *International Conference on Raman and Luminescence Spectroscopy in Technology*
Conference 823, *Optical Materials Technology for Energy Efficiency and Solar Energy Conversion VI.*

Program Chair: **Solomon Musikant,** General Electric Company

ADVANCES IN NONLINEAR POLYMERS AND INORGANIC CRYSTALS,
LIQUID CRYSTALS, AND LASER MEDIA

Volume 824

INTRODUCTION

As in all engineering creations, materials limit optical engineering applications. In earlier times the discovery of new natural materials, such as birefringent calcite, led to new physical understanding of optical phenomena and ultimately to new and useful devices. In more modern times, material scientists in conjunction with optical scientists and engineers have purposefully sought new materials to perform specific tasks that known materials cannot accomplish either at all or perhaps imperfectly. The classical example of the development of new optical materials is the work of Schott in the late 1800s in the discovery and development of new families of optical glasses. His work was a watershed in optical engineering history.

Since 1981, SPIE has pioneered in bringing together leading materials scientists working on advanced optical materials and has documented this work in a series of symposia proceedings and *Optical Engineering* journal articles. This editor has been privileged to work in this arena by being chair of seven international conferences as well as materials program chair at SPIE's annual symposia in San Diego in 1987 and 1988 (upcoming) and the editor of a special issue of *Optical Engineering* journal on optical materials (Feb. 1987). As a result of these conferences and symposia approximately 300 papers have been published, representing the greatest collection of literature on optical materials available from one source to the scientific community.

The current conference is part of this series of meetings. Although covering a relatively broad topic list, this meeting was dominated by papers in the area of nonlinear polymers under the leadership of my cochair, Dr. Garo Khanarian. The balance of the papers in this conference was divided almost equally between liquid crystal and laser media topics.

Signifying the intense activity in this area at the current time, a fair number of papers on nonlinear polymers were contributed spontaneously without the usual solicitations that are required to develop a program. The field is of great interest not only to the scientific community but to the commercial world, since many devices can be conceived of using these films and leading to mass-produced, economical, and compact instruments and appliances. Of the 17 papers on nonlinear optical polymers, approximately one third of the papers deal with applications despite the early stage of development of the technology.

Solomon Musikant
General Electric Company

ADVANCES IN NONLINEAR POLYMERS AND INORGANIC CRYSTALS,
LIQUID CRYSTALS, AND LASER MEDIA

Volume 824

Session 1

Liquid Crystals and Laser Media

Chair
William A. Penn
General Electric Company

Nonlinear optical interactions in low-melting-temperature glasses containing organic dyes

Wayne R. Tompkin and Robert W. Boyd

Institute of Optics, University of Rochester, Rochester, NY 14627

ABSTRACT

We present the results of an experimental investigation of the nonlinear optical properties of boric-acid glass doped with the organic dye fluorescein. This material has a very large third-order nonlinear optical susceptibility of approximately 1 esu, which is shown to be electronic and not thermal in origin. Measurements of the tensor nature of the nonlinear susceptibility are presented.

1. INTRODUCTION

Solid glass matrices containing organic dye molecules have been shown to possess highly desirable nonlinear optical properties. We have previously studied the organic dye fluorescein doped into boric-acid glass and have shown this material to have a very small saturation intensity of ~15 mW/cm^2, a response time of ~0.1 sec, and a $\chi^{(3)}$ susceptibility as large as 1 esu.[1] The present paper describes the results of further investigations of this material; specifically, we present evidence that the nonlinearity is electronic and not thermal in origin and we examine the tensor properties of the susceptibility.

2. SAMPLE PREPARATION AND PROPERTIES

We have used the method of Todorov et al[2] to prepare 100-μm-thick samples of boric-acid glass doped to a concentration of approximately 10^{-3} M with fluorescein having a molecular weight of 332.32. The absorption of the doped glass had a peak at 438 nm and a linewidth (FWHM) of 43 nm, making this glass particularly convenient for use with an argon-ion laser. The small-signal absorption length ($\alpha_o \ell$) of the samples was controlled by varying both the thickness of the sample and the concentration of the dye.

3. PREVIOUS WORK

Fluorescein-doped boric acid glass has been shown to display interesting and useful nonlinear optical properties.[1,2] It has been demonstrated that it is possible to perform phase conjugation by degenerate four-wave mixing (DFWM) and two-beam coupling using unfocused continuous-wave laser beams. The results of these experiments indicate a response time τ equal to 100 ms and a nonlinear susceptibility $\chi^{(3)}$ equal to 0.04(1+0.2i) esu at 488 nm as measured by two-beam coupling and $\chi^{(3)}$ equal to 0.05(1+0.3i) esu at 476.5 nm as measured using phase conjugation. The nonlinear optical properties have been shown to depend on the polarization state of the saturating field due to the fact that the individual molecules are not free to rotate. From the polarization-dependent saturation experiments it was found that the saturation intensity I_s is equal to 11 mW/cm^2 and that the susceptibility $\chi^{(3)}$ is equal to 2(1+i) esu at 457.9 nm for a sample with $\alpha_o \ell = 4.76$. The variation in the values of the nonlinear susceptibility $\chi^{(3)}$ are due presumably to differences in laser detuning and to variations in the the fluorescein concentration.

4. CONSIDERATION OF THERMAL EFFECTS

The origin of the extremely large nonlinearity displayed by fluorescein-doped boric acid glass is the long excited state lifetime[3] and the concomitant low saturation intensity. We have previously presented evidence that the origin of the nonlinearity is saturated absorption and is not a thermal effect. This evidence entailed measuring the nonlinear susceptibility using samples of boric-acid glass containing Rhodamin 6G instead of fluorescein. These new samples have the same thermal characteristics as fluorescein-doped boric acid glass, but display only prompt fluorescence and not delayed fluorescence as is observed in the case of the fluorescein dopant. These new samples produced no measurable four-wave mixing signal. This result indicates that the thermal effects play a relatively insignificant role in the nonlinear interactions.

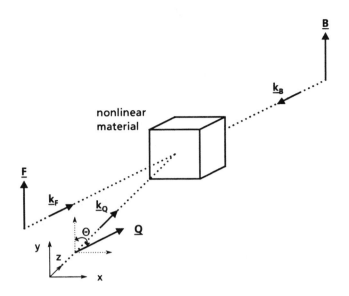

Fig. 1. Optical arrangement for the degenerate four-wave mixing experiment described in the text.

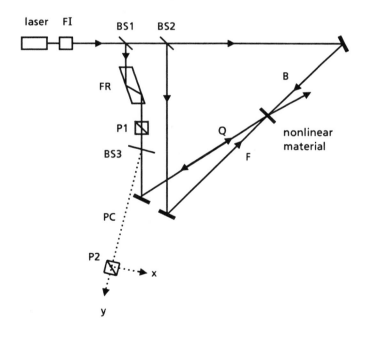

Fig. 2. Experimental arrangement for degenerate four-wave mixing experiments. A Faraday isolator (FI) precludes feedback into the argon-ion laser. BS1 is a 4% beamsplitter; BS2 and BS3 are 50/50 beamsplitters. The polarization of the probe is controlled by a linear polarizer (P1) following a Fresnel rhomb (FR). The polarization of the phase-conjugate signal (PC) is determined with an analyzer (P2).

To investigate further the origin of the large nonlinear coupling, we have measured the grating decay time as a function of the angle between the forward-going pump wave and the probe wave. These results show that the grating decay time is essentially constant as the angle between the beams is increased from 0 and 90 degrees; the measured decay time was 84 ± 5 ms. If the origin of the nonlinearity were thermal, the grating decay time would vary proportionally to the square of the grating period.[4] In addition, we have studied the phase-conjugate reflectivity as a function of the angle between the forward-going pump beam and the probe and have shown the phase-conjugate reflectivity to be essentially constant, having a standard deviation of less than 2.5% of the mean reflectivity, for all crossing angles. We conclude that thermal effects play at most a minor role in the nonlinear interactions studied.

5. VECTOR PHASE CONJUGATION

Phase-conjugate mirrors have the ability to correct for wavefront aberrations; however, it is desirable that phase-conjugate mirrors also be able to remove the effects of polarization aberrations. A mirror having this unique and desirable property is called a vector phase conjugate mirror. In order to assess the ability of DFWM in fluorescein-doped boric acid glass to perform vector phase conjugation, we have studied some of the tensor properties of the third-order nonlinear susceptibility of this material. Maker and Terhune[5] have shown that the induced third-order polarization for an isotropic material can be written as

$$\underline{P}^{NL} = a(\underline{E}\cdot\underline{E}^*)\underline{E} + \frac{1}{2}b(\underline{E}\cdot\underline{E})\underline{E}^* \qquad (1)$$

where $a=3(\chi_{1122}+\chi_{1212})$ and $b=6\chi_{1221}$. We have studied the induced polarization using DFWM in the geometry shown in Fig. 1; the total electric field $\underline{E}$ is then

$$\underline{E} = \underline{F} + \underline{B} + \underline{Q} \qquad (2)$$

where $\underline{F}$, $\underline{B}$, and $\underline{Q}$ denote the electric field vectors of the forward-going pump wave, the backward-going pump wave, and the probe wave, respectively. If one considers only the phase-matched terms, one can rewrite Eqs. (1) and (2) in matrix form as

$$\begin{bmatrix} P_x \\ P_y \end{bmatrix} = \begin{bmatrix} (2a+b)B_xF_x + bB_yF_y & a(B_xF_y+B_yF_x) \\ a(B_xF_y + B_yF_x) & (2a+b)B_yF_y + bB_xF_x \end{bmatrix} \begin{bmatrix} Q_x^* \\ Q_y^* \end{bmatrix}, \qquad (3)$$

as was shown by Saikan and Kiguchi.[6] Note that for $a\ll b$ the 2×2 matrix will be a multiple of the identity matrix, and the nonlinear polarization will be proportional to the complex conjugate of the probe field vector. Hence, for $a\ll b$ perfect vector phase conjugation will be obtained for any polarizations of the pump waves. Consider now the case where both pump waves are linearly polarized along y, that is, $\underline{B}=B\hat{y}$ and $\underline{F}=F\hat{y}$, and the probe is linearly polarized at an angle θ with respect to the pumps, that is, $\underline{Q}=Q(\sin\theta\ \underline{\hat{x}}+\cos\theta\ \hat{y})$. The expression for the nonlinear polarization, Eq. (3) then becomes

$$\begin{bmatrix} P_x \\ P_y \end{bmatrix} = QBF \begin{bmatrix} b & 0 \\ 0 & 2a+b \end{bmatrix} \begin{bmatrix} \sin\theta \\ \cos\theta \end{bmatrix}, \qquad (4)$$

Note that for $a\ll b$ the induced third-order polarization is parallel to the probe polarization.

The electric field of the signal in the ξ,η-plane a distance d from the nonlinear material is given by

$$\underline{E}_{sig}(\xi,\eta) = \frac{4\pi^2}{\lambda^2} \int dx\ dy\ dz\ \frac{\underline{P}^{NL}(x,y,z)}{r}\ e^{-ikr} \qquad (5)$$

where the integral is taken over the volume of the nonlinear material and where $r=[(x-\xi)^2+(y-\eta)^2+d^2]^{1/2}$. If we assume that the induced polarization is constant throughout the nonlinear material, then in the Fresnel approximation the electric field is given by

$$\underline{E}_{sig} = \frac{2\pi^2 \ell}{\lambda^2} \, \underline{P}^{NL} e^{ikd} \tag{6}$$

where ℓ is the thickness of the nonlinear material and $\underline{P}^{NL}$ is found from Eq. (4). The intensities associated with the x and y components of the electric field of the signal are then given by

$$I_x = \frac{2\pi^3}{\lambda^2} n_o c \ell^2 |\underline{P}^{NL} \cdot \hat{\underline{x}}|^2 = K \, |b|^2 \sin^2\theta \tag{7a}$$

and

$$I_y = \frac{2\pi^3}{\lambda^2} n_o c \ell^2 |\underline{P}^{NL} \cdot \hat{\underline{y}}|^2 = K \, |2a+b|^2 \cos^2\theta \tag{7b}$$

where $K = (2\pi^3/\lambda^2) n_o c \ell^2 (QBF)^2$ and n_o is the linear refractive index at a wavelength λ.

In order to determine the ratio $|b/a|$ from the measurable quantities I_x and I_y, we make an assumption about the ratio of the real and imaginary parts of a and b. The theoretical expression for the third-order nonlinear susceptibility $\chi^{(3)}$ of a saturable absorber is given by

$$\chi^{(3)} = -\frac{n_o^2 c^2 \alpha_o}{4\pi^2 \omega I_s} (\delta + i) \tag{8}$$

where α_o is the unsaturated absorption coefficient, I_s is the saturation intensity, and $\delta = (\omega - \omega_o) T_2$ is the detuning of the laser frequency ω from the resonance frequency ω_o, normalized to the dipole dephasing time T_2. We then conclude that, since $a = 3(\chi_{1122} + \chi_{1212})$ and $b = 6\chi_{1221}$, we can write $a = a_o(\delta + i)$ and $b = b_o(\delta + i)$ where a_o and b_o are real. With this assumption, one can solve for $|b/a|$:

$$\frac{|b|}{|a|} = \frac{b_o}{a_o} = \frac{2}{\sqrt{I_y/I_x}\tan\theta - 1} \, . \tag{9}$$

Using the experimental arrangement shown in Fig. 2, we have measured the intensities of the phase-conjugate signal polarized along the x and y directions as functions of the angle θ between the polarization of the probe and the polarization of the pumps. The polarization state of the probe was controlled by rotating a polarizer following a Fresnel rhomb. The results of such measurements for three intensities are shown in Fig. 3. The intensities of the x and y components of the signal are fitted to the predictions of Eqs. (7) using a least-squares algorithm. In the case of Fig. 3a, the two intensities are described best by $I_x = (0.010 \text{ mW/cm}^2)\sin^2\theta$ and $I_y = (0.035 \text{ mW/cm}^2)\cos^2\theta$; using Eq. (9), one then finds that $|b/a| = 2.4$. This value for $|b/a|$ gives a quantitative description of the ability of the material in correcting for polarization aberrations. We can also see from Fig. 3 that the value of $|b/a|$ increases with increasing laser intensity; hence the ability of the system to correct for polarization aberrations increases with laser intensity. The reason for the variation of $|b/a|$ with intensity appears to be that the coefficients a and b saturate with increasing laser intensity at different rates.

We have also determined the magnitudes of a, b, and $\chi^{(3)}$ from the measurements. From Eq. (7a) and the data from Fig. 3a we find that $|b| = 4\times10^{-3}$ cm^3/erg. Similarly, from Eq. (4) we find that $|\chi^{(3)}| = 10^{-3}$ cm^3/erg.

6. SUMMARY

We have presented evidence that the origin of the nonlinearity of fluorescein-doped boric acid glass is saturated absorption and not a thermal effect. The grating decay time and the DFWM reflectivity are found to be constant as a function of the angle between the pump and probe waves; these observations suggest the usefulness of this material in phase

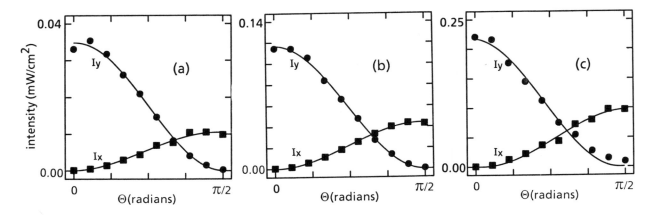

Fig. 3. The intensities of the x and y components of the electric field of the phase-conjugate signal plotted as functions of the angle between polarization of the probe and the polarization of the pumps for three intensities. For all three cases shown, the wavelength λ is 476.5 nm, the linear refractive index n_o is 1.5, the length of the non-linear material ℓ is 0.004 cm, the pumps are polarized parallel to the y-direction, the small-signal absorption coefficient α_o is 77.5 cm^{-1}, and the intensity of the probe is 6% that of the pump intensity I_p. The parameters for the three cases are: (a) I_p=0.47 W/cm^2, I_x=(0.010 mW/cm^2)sin$^2\theta$, I_y=(0.035 mW/cm^2)cos$^2\theta$, and $|b/a|$=2.4; (b) I_p=1.06 W/cm^2, I_x=(0.044 mW/cm^2)sin$^2\theta$, I_y=(0.116 mW/cm^2)cos$^2\theta$, and $|b/a|$=3.2; and (c) I_p=2.3 W/cm^2, I_x=(0.096 mW/cm^2)sin$^2\theta$, I_y=(0.218 mW/cm^2)cos$^2\theta$, and $|b/a|$=4.1.

conjugation systems utilizing a large field of view. We have also examined the ability of the system to correct for polarization aberrations. We have found that its ability to act as a vector phase conjugate mirror is enhanced through the use of saturating pump waves.

7. ACKNOWLEDGMENTS

We gratefully acknowledge useful discussions of this work with D. J. Gauthier, M. A. Kramer, K. R. MacDonald, and M. S. Malcuit. This work was supported by the Office of Naval Research.

REFERENCES

1. M.A. Kramer, W.R. Tompkin, and R.W. Boyd, "Nonlinear optical properties of fluorescein in boric-acid glass," J. Lumin. 31/32, 789 (1984); "Nonlinear-optical interactions in fluorescein-doped boric acid glass," Phys. Rev. A 34, 2026 (1986).
2. T. Todorov, L. Nikolova, N. Tomova, V. Dragostiva, "Photochromism and dynamic holographic recording in a rigid solution of fluorescein," Opt. Quantum Electron. 13, 209 (1981).
3. G.N. Lewis, D. Lipkin, and T.T. Magel, "Reversible photochemical processes in rigid media. A study of the phosphorescent state," J. Am. Chem. Soc. 63, 3005 (1941).
4. M.H. Garrett and H.J. Hoffman, "Thermally induced phase-conjugation efficiency and beam-quality studies," J. Opt. Soc. Am. 73, 617 (1983).
5. P.D. Maker, R.W. Terhune, and C.M. Savage, "Intensity-dependent changes in the refractive index of liquids," Phys. Rev. Lett. 12, 507 (1964); P.D. Maker and R.W. Terhune, "Study of optical effects due to an induced polarization third order in the electric field strength," Phys. Rev. 3A, 801 (1965).
6. S. Saikan and M. Kiguchi, "Generation of phase-conjugated vector wave fronts in atomic vapors," Opt, Lett. 7, 555 (1982).

Nd:YAG Laser with Cholesteric Liquid Crystal Cavity Mirrors

Jae-Cheul Lee, Stephen D. Jacobs and Rachel J. Gingold

University of Rochester
Laboratory for Laser Energetics
250 East River Road
Rochester, New York 14623-1299

Abstract

Lasing mechanisms for a laser resonator with cholesteric liquid crystal end mirrors were investigated. The effective radius of curvature of a cholesteric liquid crystal mirror was measured. The slope efficiency of a CLC–dielectric resonator was measured and compared with a flat-flat dielectric resonator with or without a pinhole for a Gaussian output.

1. Introduction

Cholesteric liquid crystals (CLC) have been used as polarizers, bandpass and notch filters, mirrors, apodizers and optical isolators.[1,2] These are all based on the linear propagation of light through the medium. In this case, the changes of helical structure of the CLC due to an optical field are almost negligible. But in intense optical fields, the helical structure can be changed and nonlinear effects can occur. In our previous work,[3] we showed that under exposure to a plane wave with a Gaussian intensity distribution, a retro-self focusing effect occurred in which the reflected field comes to a focus. This was accompanied by a pinholing effect in which the retro-self-focusing occurs within $\sqrt{2}w$, where $2w$ is the spot size of an incident Gaussian beam. As a result, when a CLC was used as a resonator mirror, TEM_{oo} mode operation was obtained.

In 1980, Denison and co-workers[4] reported the operation of a CuI vapor laser with CLC cavity end mirrors. Temperature tuning for operation at 510.6 nm and 578.2 nm was demonstrated, but no characterization of the output beam and no detailed lasing mechanism were provided.

In this paper, we will discuss the lasing mechanism for the CLC resonator, the cause for its alignment simplicity and its angular insensitivity. We will describe our measurement of the effective radius of curvature of the CLC mirror. The slope efficiency of the CLC-dielectric resonator will be compared with one of a flat-flat dielectric resonator.

2. Theory

Let us consider a right-handed CLC mirror as shown in Fig. 1(A) which has a planar structure. Here the CLC has a strong surface anchoring at $z = 0$ and no surface anchoring at $z = L$ (where L is the CLC fluid thickness). When a plane wave with right handed circular (RHC) polarization is incident on this CLC mirror, the relationship between incident (I) and transmitted field intensity (J) is:[5]

$$I = u_3 + \frac{u_2 - u_3}{1 - (u_1 - u_2)(u_1 - u_3)^{-1} Sn^2 (2\kappa L/g, k)} \tag{1}$$

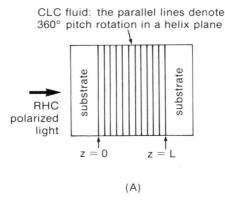 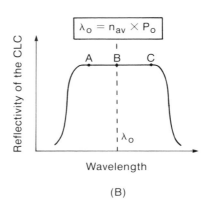

CLC fluid: the parallel lines denote 360° pitch rotation in a helix plane

RHC polarized light

substrate

$z = 0$

substrate

$z = L$

$\lambda_o = n_{av} \times P_o$

A   B   C

Reflectivity of the CLC

λ_o

Wavelength

(A)

(B)

Fig. 1 (A)   Schematic diagram of a cholesteric liquid crystal.
(B)   Its selective reflection in a planar structure, where A corresponds to the case $\Delta k/\kappa > 0$; B, $\Delta k/\kappa = 0$; and C, $\Delta k/\kappa < 0$.

where $u_1 > u_2(=J) > u_3 > u_4$ are real roots of

$$Q(u) \; = \; (u - J) \; \{u - (u - J) \; (\Delta k/\kappa + u - J)^2\} \tag{2}$$

where $\Delta k = (k_o^2 - q_o)^2/2q_o$, $k_o^2 = (\omega/c)^2 \, (\varepsilon_\parallel + \varepsilon_\perp)/2$, $\kappa = (\omega/c)^2 \, \varepsilon_a/4q_o$ and Sn is a Jacobian elliptic function with

$g = 2/[(u_1-u_3)(u_2-u_4)]^{1/2}$, $k = [(u_1-u_2)(u_3-u_4)]^{1/2}g/2$, where $\varepsilon_\parallel$ and $\varepsilon_\perp$ represent the dielectric constant parallel and perpendicular to the local director, $\varepsilon_a = \varepsilon_\parallel - \varepsilon_\perp$ is the optical dielectric anisotropy and q_o is the unperturbed wave number of the helix whose pitch is $p_o=2\pi/q_o$. The intensities I, J are normalized with $\gamma = \varepsilon_a/32\pi \, k_{22}\kappa^2$ where k_{22} is a Frank constant for twist. The relationship between incident field intensity I and transmitted field intensity J for $\kappa L = 2.6$ with the detuning parameter $\Delta k/\kappa = 0, \pm 0.1$ is shown in Fig. 2(A). This figure shows the optical bistability as theoretically predicted by H. G. Winful.[5] Physically $\Delta k/\kappa = 0$ means that the pitch of the CLC is exactly Bragg-matched to the wavelength of the incident field by the Bragg reflection condition $\lambda_o = n_{av} \cdot p_o$ where λ_o is the wavelength of selective reflection and n_{av} average refractive index for the liquid crystal. $\Delta k/\kappa < 0$ means that the wavelength of the incident optical field is longer than the selective peak wavelength λ_o, and $\Delta k/\kappa > 0$ describes the reverse condition. In Fig. 1(B), $\Delta k/\kappa = 0$ corresponds to the position B and $\Delta k/\kappa < 0 (>0)$ corresponds to the position C (and A).

The reflectivity R of a CLC mirror can be defined by

$$R \; = \; \frac{\text{reflected field intensity}}{\text{transmitted field intensity}} = 1 - J/I \tag{3}$$

A plot of reflectivity as a function of the normalized intensity I and of the detuning parameter $\Delta k/\kappa = 0, \pm 0.1$ for $\kappa L = 2.6$ is shown in Fig. 2(B). Note that over a large normalized intensity region, this reflectivity is constant. This characteristic opens the opportunity for CLC cells as laser end mirrors.

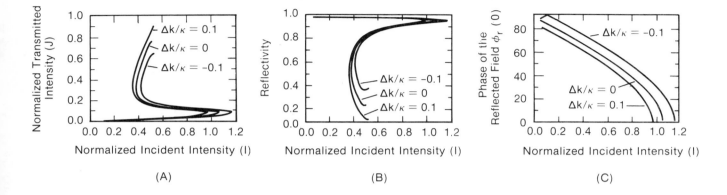

Fig. 2 (A) Transmitted intensity vs incident intensity for $\kappa L = 2.6$ for various values of the detuning parameter $\Delta k/\kappa = 0$, ± 0.1. Intensities are normalized by $8 = 32\pi k_{22}\kappa^2/\varepsilon_a$.

(B) Incident field reflectivity vs intensity.

(C) The phase of the reflected field vs incident field intensity when the phase of the incident field $\phi_i(0) = 0$.

The phase $\phi_r(z)$ of the reflected field at $z = 0$ can be written as:[4]

$$\phi_r(0) = \phi_i(0) - \cos^{-1}[-\sqrt{1-J/I}\{\Delta k/\kappa + (I-J)\}] \qquad (4)$$

where $\phi_i(0)$ is the phase of the incident field at $z = 0$. Equation (4) shows that the phase of the reflected field depends on the field intensity. For $\phi_i(0) = 0$, a plot of the phase $\phi_r(0)$ as a function of the normalized input intensity I and the detuning parameter $\Delta k/\kappa$ is shown in Fig. 2(C). Note that for $\Delta k/\kappa = 0$, the phase ϕ_r initially starts from 90°. When $\Delta k/\kappa = 0$, that is, the CLC structure is initially well–Bragg matched, the intense field destroys the Bragg condition and the optical field penetrates more deeply into the CLC before being reflected. Therefore the phase of the reflected field lags as the input intensity increases. When $\Delta k/\kappa > 0$, the pitch of the CLC is already out of the Bragg condition. So the initial phase is less than 90° and decreases further with increasing intensity. When $\Delta k/\kappa < 0$, the incident field forces a pitch dilation, and the CLC will approach the Bragg-condition. Therefore the phase increases until the Bragg-condition is met and then decreases.

Now let us consider a plane wave with a Gaussian intensity distribution incident on the CLC. In a first approximation, we apply our plane wave consideration to study the transverse effects. In Fig. 1(B), when $\Delta k/\kappa \geq 0$ [position B and A] and the maximum pitch dilation is restricted to the selective reflection band, the phase of the reflected field lags as the intensity increases. Therefore, the CLC acts as a concave mirror causing a retro-self-focusing effect. When $\Delta k/\kappa < 0$ [position C], the maximum pitch dilation may drive point B past position C whereupon the CLC becomes a convex mirror. If the maximum pitch dilation crosses over the position B, the CLC acts as a combination of concave and convex mirrors. This situation is described in Fig. 3. The curvature of the CLC mirror depends on the location of the selective reflection peak wavelength λ_o relative to the laser wavelength λ and on the maximum intensity of the incident field. For laser end mirrors, the condition $\lambda_{laser} \leq \lambda_o$ is important because the CLC can be modeled as a concave mirror in this region. When it is used as a laser end mirror, it becomes an active element which changes its radius of curvature in proportion to the intensity.

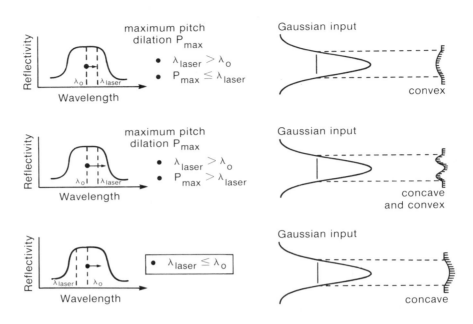

Fig. 3   The transverse curvature modulation of the CLC mirror by the incident Gaussian intensity distribution.

The lasing mechanism of a CLC resonator is as follows: initially the entire spontaneous emission spectrum from the active medium within the selective reflection band is involved in the pitch dilation. But since the spatial intensity distribution of the spontaneous emission is bell-shaped, the CLC acts as a well aligned curved mirror and the beam reflects back to the rod and eventually stimulated emission occurs. For this reason the CLC resonator is easy to align, at a long mirror separation. Once above lasing threshold, the strong intra-cavity field reorients the molecular structure of the CLC mirror even if the substrate is being tilted. At this point, if the CLC resonator is turned off and on again it does not resume lasing. Therefore, some adiabatic following of the molecular structure with strong fields occurs. This explains the reason why the CLC resonator has angular insensitivity.[3]

3. Sample Preparation

Several 1.5-inch diameter CLC mirrors were assembled using BK–7 substrates. The mirror substrates were separated by mylar spacers of a known thickness. The gap was then filled by capillary action with a right-handed cholesteric liquid crystal mixture of CB15 and E7. For strong surface anchoring, one side of the substrate was coated with a solution of 0.2% polyvinyl alcohol (PVA) in water by a spinning process, placed on a hot plate at 120°C for 15 minutes, and then cooled down gradually to room temperature. The PVA coated surface was lightly buffed with a short-fiber paint roller rotating on a level platform, to give the liquid crystal molecules an alignment direction.

Three different kinds of CLC mirrors were made with different anchoring. Here we will introduce a convenient identification code to describe the CLC mirrors. The format is:

$$XX\#\# - \#\#$$

The first two characters XX indicate the handedness of the helix structure: RH (right-handed helix structure) or LH (left-handed helix structure). The next two characters ## indicate the fluid thickness in units of μm and the last two characters describe the state of anchoring,

00 : no surface anchoring at both sides.
01 : no anchoring at z = 0, strong anchoring at z = L.
10 : strong anchoring at z = 0, no anchoring z = L.
11 : strong anchoring at both sides.

For example, RH10–10 indicates a CLC mirror with a right-handed helix structure, 10-μm fluid thickness and a strong anchoring at z = 0. This CLC can also be designated as RH10–01 if the cell is flipped end-for-end in the incident beam.

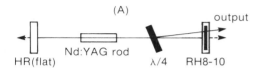

(A)

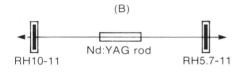

(B)

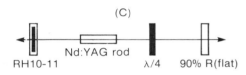

(C)

Fig. 4 The CLC resonator configuration tested in the experiment.

(A) CLC (thick) - dielectric (flat) resonator: output via Fresnel reflection off the slightly tilted waveplate.

(B) CLC (thick) - CLC (thin) resonator: output through thin partial reflector.

(C) CLC (thick) - dielectric resonator: output through 90% R output coupler.

4. Experiment

A commercial Nd:YAG laser (Control Laser Corp. Model 256) has been employed for this work. The resonator configurations considered are shown in Fig. 4. In Fig. 4(A), the CLC–dielectric resonator consists of a flat dielectric HR end mirror, a Kr arc-lamp pump, 54-mm long by 4 mm diameter Nd:YAG laser rod, an uncoated λ/4 plate, and a CLC RH8-10 end mirror. Output beam patterns were taken with IR film and a CID[6] camera at both ends of the resonator, respectively. Figure 5 shows the output beam pattern leaked through the HR side. Data captured by the CID camera were passed to a beam analyzer which consists of a PDP–11/73 and beam analysis software. Contouring patterns show the beam symmetry, and a Gaussian fit of the output intensity distribution shows that a perfect Gaussian beam was generated inside the resonator as a result of pinholing effects. The ellipticity was about 1.04. The output patterns at the CLC mirror side are shown in Fig. 6. These photographs were taken with IR film approximately 4 meters from the output mirror. The 180° phase shift upon Fresnel reflection from the λ/4 plate permits output coupling through the CLC mirror. The coupled output beam is shown in Fig. 6(A). A quadrupole scattering pattern due to the nonuniform helical structure change was also observed as shown in Fig. 6(A). This scattering pattern was generated in the RH CLC mirror by the RHC polarized light inside the resonator. It was similar to the calculated scattering pattern for a two dimensional anisotropic disk.[7] At higher lamp current (~16A), a ring pattern evolved with an intensity dip at the center and a scattering pattern as shown in Fig. 6(B). A more detailed mechanism will be described elsewhere.

The anchoring effect on the CLC-dielectric resonator in Fig. 4(A) was investigated. Three different kinds of CLC mirrors, that is, RH10–00, RH10–01 and RH10–11 were used. The resonator lased relatively easily with RH10–10 and RH10–11 CLC mirrors. Lasing could not be obtained with RH10–00 and RH10–01 mirrors because the lack of surface anchoring at z=0 prevented pitch dilation from occurring. In any case, the degree of uniformity in anchoring and CLC structure orientation throughout the bulk is critically important for a high quality output. A lack of this property causes the spurious scattering.

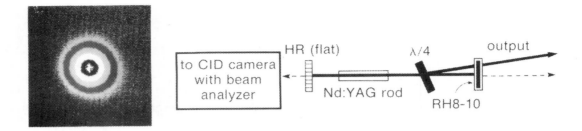

● Sampling of leakage through HR shows perfect Gaussian in resonator.

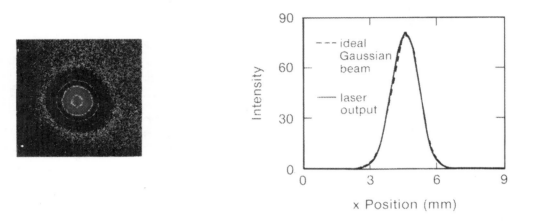

Fig. 5   Resonator configuration A: Quality of output. Computer-generated intensity contour patterns and Gaussian fit to a linear scale through the 2–D contour map.

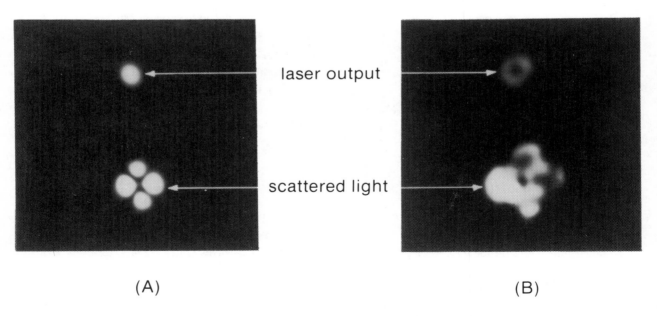

(A)

(B)

Fig. 6   The output of the CLC-dielectric resonator configuration A at the CLC mirror side (A) at low lamp current (10.5A) (B) at high lamp current (16A).

For the CLC (thick) - CLC (thin) resonator in Fig. 4(B), the output coupling through the thin CLC was demonstrated. In this configuration, the RH5.7–11 was used as an output coupler, and RH10–11 as a full reflector.

For the CLC-dielectric resonator in Fig. 4(C), retro-self-focusing effect was quantified by measuring the effective radius of curvature R of the CLC mirror. Here the YAG laser rod was assumed to be a thin lens[8] and the CLC as a concave mirror as shown in Fig. 7. In the notation of the ABCD matrix formalism widely used in laser resonator analysis, the radius of curvature, R_b, of the CLC and the spot size w_o at the output coupler are given by:[9]

$$R_b = \frac{2B}{A-D} = \infty \tag{5}$$

$$\frac{\pi \, w_o{}^2}{\lambda} = \frac{B}{\left[1 - (\frac{A+D}{2})^2\right]^{1/2}} \tag{6}$$

where

$$A = D = 1 - \frac{2d}{f_1} - \frac{d}{f_2} + \frac{3d_1d_2}{f_1f_2} + \frac{2d_1d_2}{f_1{}^2} - \frac{d_1d_2{}^2}{f_1{}^2f_2} \tag{7}$$

$$B = 2d - \frac{2(dd_1+d_1d_2)}{f_1} - \frac{d^2}{f_2} + \frac{2dd_1d_2}{f_1f_2} + \frac{2d_1{}^2d_2}{f_1{}^2} - \frac{d_1{}^2d_2{}^2}{f_1{}^2f_2} \tag{8}$$

$$d = d_1 + d_2, \quad f_2 = R/2$$

Equation (6) shows that the spot size w_o changes with radius of curvature R. If w_o and f_1 (the thermal lensing effect of the rod) are known, one can calculate the radius of curvature R. The thermal lensing effect can be measured accurately by shearing interferometry[10] without cavity end mirrors as shown in Fig. 8(A). For this experiment, a beam expander collimates a He-Ne beam which enters the laser rod. The shearing is accomplished with a shearing plate which is tilted and wedged in orthogonal directions. The interferograms generated are captured by a CID camera and displayed on the monitor. Under no optical pumping, straight fringes parallel to the horizontal direction result and these straight fringes rotate due to thermal lensing as the optical pumping power increases. The effective radius of curvature of the laser rod R_d can be estimated by

$$R_d = \frac{-st}{\lambda \sin\theta} \tag{9}$$

where R_d is the radius of curvature of the laser rod, s is the shearing distance, t is the fringe spacing and θ is the rotation angle of the straight fringes.

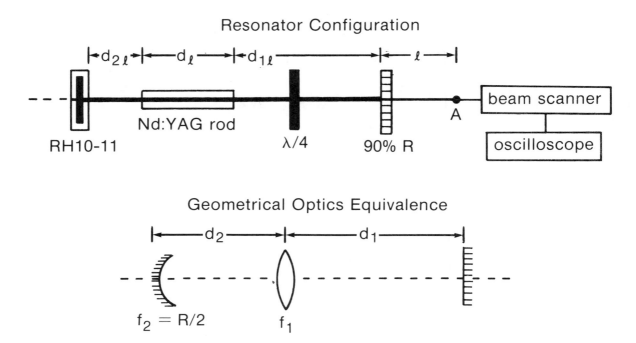

Fig. 7   Experimental set-up for measurement of the radius of curvature for CLC mirror and its geometrical optics
equivalence.

The result is depicted in Fig. 8(B). It shows that the laser rod used here acts as a negative lens below the lasing
threshold current ($\approx$ 10 A) and as a positive lens at higher current. We determined the spot size w_o by measuring the spot
size w_A at the position A [see Fig. 7] with a beam scanner[11],

$$w_o = \left[\frac{w_A^2 - \sqrt{w_A^4 - (2\lambda\ell/\pi)^2}}{2}\right]^{1/2} \tag{10}$$

where ℓ, which is equal to 286 cm, is the distance between the output coupler and position A.

We measured the thermal lensing effect of the laser rod from Fig 8(B) and calculated the spot size w_o from Eq. (10) at
two different current settings. Those results are summarized in Table I. In Fig. 7, we used $d_{1\ell}$ = 282 mm, $d_{2\ell}$ = 312 mm
and d_ℓ = 54 mm. The output intensity was confirmed to be almost Gaussian by beam-scanner derived oscilloscope traces.

As the lamp current increased, the radius of curvature of the CLC decreased from 0.9–1 m to 0.7 m. This proved the CLC
mirror to be concave and its optical power to increase with intra-cavity intensity. The cavity was in a stable region because
$d/R \leq 1$, where d is the total cavity length.

The slope efficiencies were measured for a flat-flat dielectric resonator, with and without a pinhole for Gaussian beam,
and a CLC-dielectric resonator as shown in Fig. 9. The slope efficiency for the flat-flat dielectric resonator was steepest but
multi-mode. However the output from the CLC (RH10-11)-dielectric (flat) resonator was TEM_{oo} mode with comparable
slope efficiency and threshold current. Output power was quite insensitive to mirror tilt in the CLC-resonator as compared to
the dielectric resonator. In the flat-flat dielectric resonator with a pinhole, the slope efficiency was significantly lower and its
threshold current was about 20% higher than the CLC-dielectric resonator.

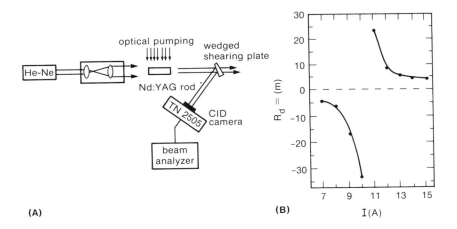

Fig. 8   (A) Experimental set-up for measurement of the thermal lensing effect of the Nd:Yag laser rod.
         (B) Effective radius of curvature of the laser rod as a function of lamp current.

Lamp Current I (A)	Spot Size W_o (mm)	Effective Radius of Curvature of		Output Beam Profile
		Laser Rod (m)	CLC Mirror (m)	
10.8	0.4-0.3	25	0.9-1.0	Gaussian
11.6	0.3	12	0.7	Gaussian

Table I   Effective radius of curvature of the CLC mirror

5. Conclusion

We have demonstrated a Nd:YAG laser with CLC end mirrors and described the performance in terms of the retro-self focusing effect and the pinholing effect. We also measured the effective radius of curvature to be ~1 m concave at moderate lamp currents. TEM$_{00}$ output of more than 3 Watts was obtained with no intra-cavity pinhole. The output profile did depend on the quality of the CLC structure. Temperature effects that are known to affect the optical performance of the CLC mirror have not been considered so far, although stable operation has been observed without any form of temperature control on the CLC mirror.

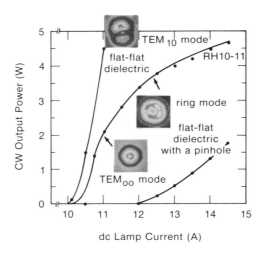

(flat-flat) dielectric resonator
- multimode output profile with no internal aperture
- high sensitivity to mirror tilt (± 0.8 arcsec)

CLC-dielectric resonator
- TEM_{00} mode operation with no internal aperture
- low sensitivity to mirror tilt (± 1.8 arcsec)

Fig. 9   The slope efficiencies for the CLC-dielectric resonator and a flat-flat dielectric resonator with or without a mode selecting intra-cavity pinhole.

Acknowledgments

The authors gratefully acknowledge A. Schmid, J. Kelly, W. Seka, K. Cerqua, T. S. Kim and J. Walker for valuable discussions, K. Marshall and K. Skerrett for help in fabricating cholesteric liquid crystals and T. Kessler and W. Castle for discussion of experiments. Special thanks are extended to Linda Boyle (manuscript preparation) and to Autumn Craft (figure preparation). This work was supported in part by Daewoo Heavy Industries, Ltd., of Incheon, Korea, the U.S. Department of Energy Office of Inertial Fusion under agreement No. DE-FC08-85DP40200 and by the Laser Fusion Feasibility Project at the Laboratory for Laser Energetics which has the following sponsors: Empire State Electric Energy Research Corporation, General Electric Company, New York State Energy Research and Development Authority, Ontario Hydro, and the University of Rochester. This work was also supported in part by the New York State Center for Advanced Optical Technology of the Institute of Optics. Such support does not imply endorsement of the content by any of the above parties.

References

1.   S. D. Jacobs, "Liquid Crystals for Laser Applications," Optical Materials - Properties, edited by M. J. Weber, Part 2, (CRC Press, Boca Raton, FL, 1986), pp. 409-465.
2.   S. D. Jacobs, K. A. Cerqua, K. M. Marshall, T. J. Kessler, R. J. Gingold, P. J. Laverty, M. Topp, "High Power Laser Beam Apodization Using a Liquid Crystal Soft Aperture," Technical Digest - CLEO '86, THG2, pp. 258-259, San Francisco, CA (June 1986).
3.   Jae-Cheul Lee, S. D. Jacobs and A. Schmid, "Retro-Self-Focusing and Pinholing Effect in a Cholesteric Liquid Crystal," submitted to Molecular Crystals and Liquid Crystals incorporating Nonlinear Optics.

4.   Yu. V. Denisov, V. A. Kizel', V. A. Orlov, and N. F. Perevozchikov, "Properties of radiation emitted by a laser with liquid-crystal reflectors," Sov. J. Quant. Electron. 10, 1447-1448 (1980).

5.   H. G. Winful, "Non-linear Reflection in Cholesteric Liquid Crystals: Mirrorless Optical Bistability," Phys. Rev. Lett. 49, 1179-1182 (1982).

6.   GE model TN 2505, array size: 248 V x 388 H pixels, pixel size: 27.3 μ x 23.3 μm.

7.   M. B. Rhodes, R. S. Porter, W. Chu, and R. S. Stein, "Light Scattering Studies of Orientation Correlations in Cholesteryl Esters," Molecular Crystals and Liquid Crystals 10, 295-316 (1970).

8.   Vittorio Magni, "Resonators for solid-state lasers with large-volume fundamental mode and high alignment stability," Appl. Opt. 25, 107-117 (1986).

9.   Joseph T. Verdeyen, Laser Electronics (Prentice-Hall, Englewood Cliffs, NJ, 1981), pp. 94-95.

10.   Daniel Malacara, Optical Shop Testing (Wiley, New York, 1978), pp. 105-111.

11.   Beam Scan, Model 1086-16, Photon, Inc.

Infrared Optical Wave Mixing and Beam Amplification with Liquid Crystal Nonlinearity

I. C. Khoo, G. M. Finn, T. H. Liu, P. Y. Yan and R. R. Michael
Electrical Engineering Department
The Pennsylvania State University
University Park, PA 16802

Abstract

We present a detailed theory and experimental study of infrared optical wave mixing based on thermal nonlinearity in nematic liquid crystal films. Because of the longer wavelength of infrared laser, lower scattering loss, transparency and other unique physical characteristics, very efficient degenerate optical wave mixing effects can be realized in nematic liquid crystal films. Applications to fairly fast (submillisecond) phase conjugations, beam and image amplification and self-pumped phase conjugation are also discussed.

Introduction

Current research and development in optical imaging, adaptive optics, communication and signal processing and switching have focused considerable efforts on highly efficient nonlinear optical and electro-optical materials. In combination with novel nonlinear optical processes and cw or pulsed lasers covering a wide spectral range, several useful practical devices have been developed.[1] The progress in the research is particularly rapid in the visible and near infrared regions, where several highly nonlinear materials of unique characteristics have emerged. Materials such as photorefractive crystals, semiconductors, multiple quantum wells, liquid crystals and organic polymers have all been shown to possess rather high or fast nonlinearities that are useful for various aspects of electro-optical applications. In particular, optical phase conjugating resonators, self-induced ring oscillators, beam steering and beam combining and dynamic interferometers, optical switching elements and logic gates, infrared-to-visible image converter, real-time image correlators, image or beam amplifiers, nonlinear guided wave structures and various form of optical switches are but a few salient examples.

In great contrast to these developments in the visible and near infrared regime, there are relatively very few nonlinear optical materials further into the infrared spectrum (for example, at the 5 μm (CO laser) and the 10 μm (CO_2) spectral windows). Some semiconductors (e.g., mercury cadmium telluride[2] (Hg CdTe), Indium Antimonide[2] (InSb)), have been shown to possess very large optical nonlinearities, but are cumbersome in practical usage owing to either cryogenic cooling requirements and/or fabrication problems and cost.

This paper is focused on the special characteristics of nematic liquid crystals for highly efficient nonlinear optics applications in the infrared spectral region. In particular, nematic films are ideal for degenerate optical wave mixing involving infrared lasers. This leads to potential applications in beam and image amplifications and self-pumped phase conjugator, with modest power lasers.

Nematic liquid crystals are generally quite transparent in the infrared, possessing absorption constant on the order of 20 to 100 cm^{-1}.[3,4] They possess rather high infrared birefringence, with a Δn ($\Delta n = n_e - n_0$) of about 0.2 to 0.3. Because of the longer wavelength of infrared radiation, the scattering loss due to orientational fluctuations is also proportionally much smaller. Following the usual orientational fluctuation scattering theory, the loss decreases faster than λ^{-2}, where λ is the wavelength of light. Thus, while the scattering losses in the visible region is on the order of 20 db/cm, the corresponding loss in the 10 μm area is negligibly small, especially in the usual thin nematic film configurations. These and other physical characteristics (c.f., Reference 6) allow the extension of progresses in nonlinear optics made in the visible spectral region to the infrared regime a logical and natural progression. In this paper, we will concentrate on principally degenerate optical wave mixing; in particular, the relatively new process of beam amplifications via multiwave mixings.

Multiwave mixings:

Figure 1 depicts a typical strong pump and weak probe wave mixing set up in a nonlinear medium. As a result of the nonlinearity (orientational or thermal), a refractive index grating is set up by the incident pump and probe beams. Depending on the magnitude of the refractive index modulation, which is governed by the pump-probe beam ratio and intensity, one-, two- or more diffractive beams will occur. In a thick medium, due to phase mismatch (which depends principally on the wave mixing angle θ), these diffractive beams will be of vanishing amplitudes. However, in a highly nonlinear thin film, the diffracted beam can be of very large amplitude (compared to the weak probe) even though the probe beam is quite weak compared to the pump beam. In fact, for sufficiently weak probe beams (e.g., a pump-to-probe beam intensity ratio > 10 or more), the diffracted beam can be of the same order of magnitude as the probe beam. It is under these circumstances that the interaction between the pump beam and the diffracted beam (i.e., forming or grating) leads to substatial amplitude modulation of the probe beam. Under favorable conditions, the probe beam can be amplified via the scattering of the strong pump beam from this pump-diffracted beam grating.[7] This amplification process, which is essentially a degenerate four wave mixing process, is very different from the pump-probe two wave mixing process mediated by a phase shifted grating formed by the pump and the probe. (The phase shift maybe a naturally occurring one as in the case of photorefractive response, or one induced by moving grating technique.) The mechanism of probe amplification is actually quite a well-known one, dating back to the early days of nonlinear optics.[8] However, the emergence of very highly nonlinear materials (e.g., liquid crystal films, multiple quantum wells, excitonic nonlinearity in semiconductors, etc.) usher in new perspectives and interesting experimental possibilities.

We have recently carried out two major detailed theories for this type of wave mixing process, one involves a so-called local Kerr-like nonlinearity,[9] and the other a nonlocal (diffusive-type) thermal nonlinearity,[10] and experimental confirmations. In the former, the refractive index change is assumed to be locally (point-wise) dependent on the intensity of the optical field. This is the type of nonlinearity usually encountered or assumed in most analyses, and the resultant coupled equations involve only the field variables. (In the case where the pump and probe beam generates up to two diffracted orders, there are six coupled Maxwell wave equations that need to be solved, subject to the prevailing conditions of phase mismatches, incident intensities and other geometrical-optical initial conditions.) The medium's response is represented simply by a nonlinear coefficient n_2, i.e., the refractive index change Δn is simply related to the total optical intensity I by $\Delta n = n_2 I$, and this refractive index change serves as the driving term in the coupled Maxwell wave equations.

On the other hand, nonlocal nonlinearities (e.g., thermal effect) involve some diffusion processes. The refractive index grating is not point-wise dependent on the optical intensity distribution. Rather, it is a solution of the medium's response subject to the geometry and other physical boundary conditions, including the optical intensities. Hence, the correct solution of the refractive index change should involve a self-consistent solution of both the coupled Maxwell wave equations for the optical fields (with the intensity dependent refractive index change as the driving term), and the diffusion equation governing the mechanism for nonlinearity (with the field intensities as the driving source term). We have solved the problem explicitly for thermal gratings problem in nematic films. Details of the formalism can be found in Reference 10.

For liquid crystals, both types of nonlinearities exist. The orientational nonlinearity is Kerr-like, while the thermal nonlinearity is diffusive. In terms of faster response times, especially for grating constants (and/or film thickness) on the order of 10^2 μm, where large nonlinear responses are observed, the thermal nonlinearity is preferrable. For grating constant on the order of 200 μm, for example, the thermal decay time is on the order of a few ms($\leq$ 10 ms).

We have conducted experiments with both cw and pulsed (ms) CO_2 lasers. Figure 2 is a schematic of the optical field-nematic interaction. The nematic film is fabricated by sandwiching the liquid crystal PCB (Pentyl-Chloro-Biphenyl) between surfactant coated ZnSe window for homeotropic alignment. The film is 120 μm thick, maintained at a room-temperature of 22°C. PCB has appreciable absorption at 10.6 μm, with an absorption constant of about 80 cm^{-1}. The laser used is a cw CO_2 laser operating in the TEM_{00} mode from Advanced Kinetics. The laser is linearly polarized and is divided into a pump and a probe beam in the ratio of 60 to 1. The lasers are incident on the film with the polarization normal to the director axis of the nematic film. In this configuration, the thermal refractive index change is due to dn_0/dT, which has a value of about $10^{-3} °k^{-1}$ at 22°. The two lasers are crossed at an angle of 2-3°, producing an index grating with a grating constant of about 200 μm. At an input pump power of 0.85 Watt (laser beam diameter is 4mm), an increase in the transmitted probe beam by 10% is observed. The gain is rather nonlinearly dependent on the pump power (Figure 3). At a pump power of 1.5 watt, the gain is 40%. Thus, there is a roughly I^2 pump dependence as expected in this degenerate four wave mixing process. There are, however, several problems associated with higher cw pump power. Because of the thermal buildup, it is found that unless the cw pump laser is very precisely controlled at a certain level (below 6 watts) the sample tends to be heated into the isotropic phase. Since the thermal decay time for the geometry used (thickness ~ 120 μm; grating constant ~ 210 μm), is on the order of 10ms, it is more logical to use millisecond laser pulses than continuous illumination, as demonstrated in a previous study.[11] By choosing the right pulse length, and pulse rate, very efficient beam amplification effect can be obtained.

Figure 4a and 4b depict the results obtained with 40ms - long laser pulses (at 3 Hz) and a pump-probe ratio of 60 to 1. Figure 4a is obtained with a pump pulse energy of 40 millijoule (i.e., a pump power of 1 watt) and a probe gain of 30% is observed. On the other hand, with a pump power of 4.6 watt, a probe gain of 20 (2000 %) is observed (c.f., Figure 4b). Since we have not attempted to optimize the observed gain, we expect that even larger gain can be realized (as predicted in our theory[10]). This can be done by using thicker sample and/or larger grating constants, and the appropriate pulse duration.

The beam amplification effect as described and demonstrated above can obviously be applied in several infrared nonlinear optical devices such as ring oscillators, self-pumped phase conjugators, and infrared image amplifiers. The success of these applications depend on critical choices of several geometrical-physical-optical parameters (such as grating constant (wave mixing angle), temperature, nematics, pump-probe beam ratio, polarizations and film alignments, etc.) that optimize the specific application. These are the subjects of discussion of References 9 and 10 and some upcoming longer articles elsewhere.

This research is supported by a grant from the National Science Foundation ECS8415387.

References

1. See, for example, "Optical Phase Conjugation," ed. R. Fisher (Academic Press, NY, 1983) all chapters. See also R. F. Reintjes, "Nonlinear Optical Parametric Processes in Liquid and Gases" (Academic Press, NY, 1984).

2. See reference "Optical Phase Conjugation," chapter on semiconductors.

3. See, for example, J. G. Pasko, J. Tracy and W. Elser, SPIE Proceedings on Active Optical Devices, Vol. 202, 82 (1979).

4. See also S. T. Wu and T. D. Bates, J. Opt. Soc. Am. B3, 247 (1986) and numerous references therein.

5. P. G. deGennes, "The Physics of Liquid Crystals," (Oxford University Press, London, 1974).

6. I. C. Khoo, "Nonlinear Optics of Liquid Crystals," in "Progress in Optics," Volume XXV, ed. Emil Wolf (North Holland Press). In Press, 1988.

7. I. C. Khoo and T. H. Liu, IEEE J. Quantum Electronics, QE23, 171 (1987).

8. R. Y. Chiao, P. L. Kelley and E. Garmire, Phys. Rev. Lett. 17, 1158 (1966); R. L. Carman, R. Y. Chiao, and P. L. Kelley, Phys. Rev. Lett. 17, 1281 (1986).

9. T. H. Liu and I. C. Khoo, IEEE J. Quant. Electronics (In Press, 1988).

10. I. C. Khoo, P. Y. Yan, G. M. Finn, T. H. Liu and R. R. Michael (submitted to J. Opt. Soc. B, 1977).

11. I. C. Khoo and S. Shepard, J. Appl. Phys., 54, 5491 (1983).

Figure Captions

Fig. 1   Schematic of the coherent interaction of a strong pump and a weak probe beam in a nonlinear medium. For thin grating, several orders of diffracted beams are generated.

Fig. 2   Schematic of the interaction of linearly polarized CO_2 laser with a homeotropically aligned nematic liquid crystal film.

Fig. 3   Observed probe beam gain as a function of the input pump intensity, showing a roughly I^2pump dependence as expected in this four wave mixing process. Larger gain (>100%) is obtained if the laser is pulsed, to avoid overheating.

Fig. 4a   Oscilloscope traces of the probe beam pulse in the absence/presence of pump pulse, showing a 30% gain at 1 watt pump power.

Fig. 4b   Probe pulse gain at high pump power (4.6 watt), showing a gain of 20.

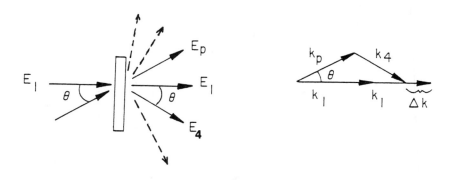

Figure 1

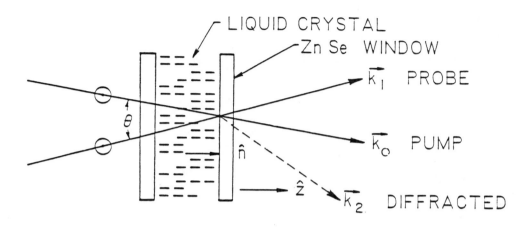

Fig.2

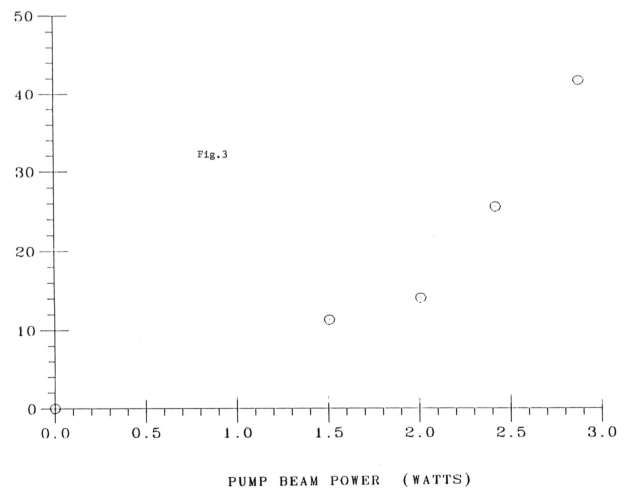

Fig.3

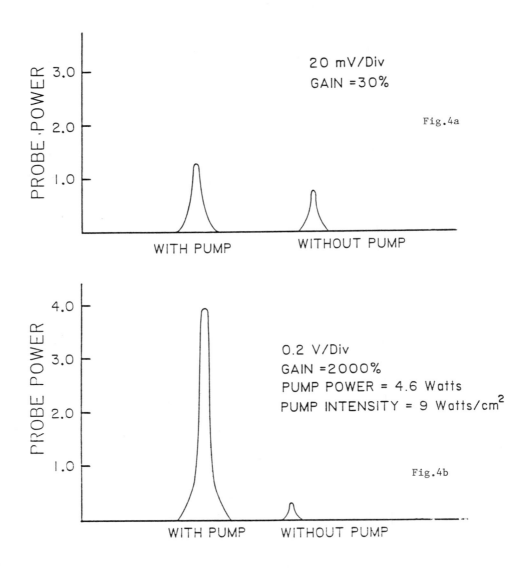

A Unified Theory for the WORM Sensitivity Function

Antonio J. Mendez

Center for Laser Studies, University of Southern California
Los Angeles, CA 90089-1112

ABSTRACT

The sensitivity function ΔR of WORM media often is a function of the incident energy (E) and pulsewidth (t), as well as incident power (P), of the write laser. A unified theory is developed to include all variables. It models the sensitivity function as $\Delta R(E,P,t) = \Delta R(Y,t)[1 - X_T^a(Y,t)/X]$, where X is the dominant and Y the secondary variable which determines media performance. $\Delta R(Y,t)$ is the saturation and $X_T^a(Y,t)$ is the threshold behavior. The theory is applied to five distinct types of organic, inorganic, and textured WORM media. The results are discussed and recommendations for test planning are made.

1. INTRODUCTION

Several properties of write once read many times (WORM) optical media can be used to describe the operational utility of the media. These include the written carrier to noise ratio (CNR), signal to noise ratio (SNR), bit error rate (BER), and the sensitivity function (delta reflectance as a function of write incident power). Although the sensitivity function is less utilized than the others mentioned, it is a very powerful descriptor of the optical media utility because it explicitly includes (a) the write threshold, (b) the saturation delta reflectance, and (c) nonlinearities and deviations from the reciprocity law. Also, it can be shown that the CNR and SNR functions are directly related to the sensitivity function.

In this paper, we will describe the sensitivity function as it is usually derived in the literature and then we will propose a modification which unifies it in the sense that more types of WORM media are covered by the modified expression, including their nonlinearities and deviations from reciprocity. The unified theory leads to a semi-empirical, closed form expression which can be used for media modelling.

2. THEORETICAL BACKGROUND

The sensitivity function was derived to describe the laser marking phenomenon at a given wavelength in terms of the background reflectance, the laser mark reflectance, the threshold for laser marking, and the saturation of laser marking [1,2]. In the referenced derivation the sensitivity function takes the form

$$\Delta R(P) = \Delta R(\infty)[1 - P_T^a/P] \tag{1}$$

where

$$\Delta R(P) = |\text{Reflectance of laser mark} - \text{reflectance of (unmarked) background}|$$

$$\Delta R(\infty) = \text{saturation } \Delta R(P) \text{ at "high" laser power incident on the media}$$

$$P_T^a = \text{apparent laser marking threshold}$$

$$P = \text{laser power incident on the media}$$

The key properties of an optical media, $\Delta R(\infty)$ and P_T^a, are readily extracted from empirical $\Delta R(P)$ data if they are plotted as a function of $1/P$, as shown in Figure 1. Then the constants of $\Delta R(P)$, that is $\Delta R(\infty)$ and P_T^a, are readily identified as the ordinate and abscissa intercepts of the $1/P$ plot, respectively. If an optical media does not really respond to P in a simple way, then the $1/P$ plot will show nonlinearities and deviations as shown in Figs. 2a through 2d. Many of these nonlinearities are caused by the fact that the laser marking phenomenon is a complex function of incident laser energy per pulse, fluence (energy per pulse per unit area), or pulse width as well as incident power. The unified theory modifies Eq (1) so that the true dependencies can be unfolded.

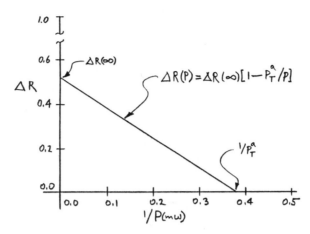

Figure 1. The 1/P plot for the sensitivity function of an ideal WORM media.

3. EXPERIMENTAL BACKGROUND

We will present here some of the experimental data which has motivated the new, unified theory. The modifications to the earlier model are based on reconciling that model with empirical observations.

The data which we analyzed includes (1) Drexler II WORM media [3,4], (2) Al-Al$_2$O$_3$ -In/InO$_x$ trilayer [5], (3) Celanese organic media [6,7], (4) hydrogen phthalocyanine media [8], and (5) textured media [9,10]. The data are shown in Figures 3a throudgh 3e. The primary observation
which we make, based on the above data, is that the WORM media sensitivity function is, in general, a complex function of incident laser power, incident energy or energy density, and pulse width.

4. RECONCILIATION OF THEORY AND EXPERIMENT

The sensitivity function or its equivalent, the reflection contrast, is typically measured and plotted as a function of incident write laser power. These plots often exhibit nonlinearities and significant data dispersion. We find that many of these nonlinearities and dispersions are removable by (1) determining whether a particular media responds primarily to energy deposited or incident laser power, (2) generating a 1/X plot, where X is the dominant independent variable, and then (3) removing nonlinearities in the domain of the secondary, or complementary, variable Y. As will be shown later, X can be P or E, then Y is E, P, or t. We find that for most WORM media, E is the dominant independent variable. Other authors have made reference to the energy density threshold for laser marking but have not used energy or energy density as the independent variable in their analyses of the sensitivity function [11,12].

We find that a sufficiently general expression which covers all experimental cases studied to date is

$$\Delta R(E,P,t) = \lim \Delta R(E \text{ or } P \text{ or } t)\,[1-(\lim X_T^a(Y))/X], \tag{2}$$

where $\Delta R(E,P,t)$ = general expression for the sensitivity function
            at a given wavelength
        E = incident laser energy or energy density
        P = incident laser power
        t = laser pulse length
  $\lim \Delta R(E \text{ or } P \text{ or } t)$
        = saturation level of R (E,P,t), written as a function of its dominant
            variable, E or P or t
        X = dominant independent variable E or P identified by experiment (e.g., by
            a power and time series of experiments)
        Y = secondary independent variable, E or P or t
  $\lim X_T^a(Y)$ = threshold value of dominant variable written as a function of the secondary
            independent variable.

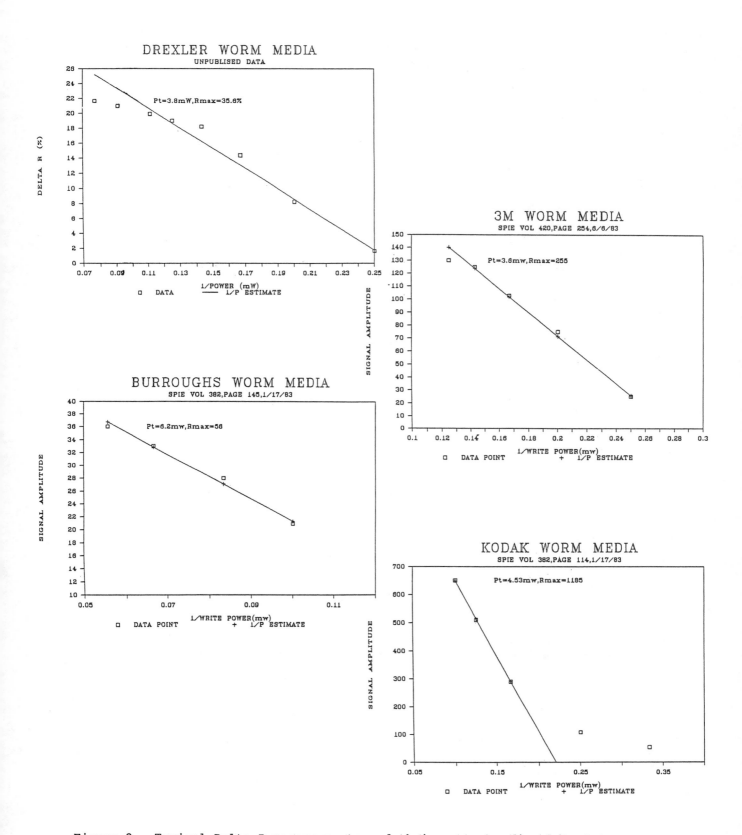

Figure 2. Typical Delta R measurements and their corresponding l/P plots.

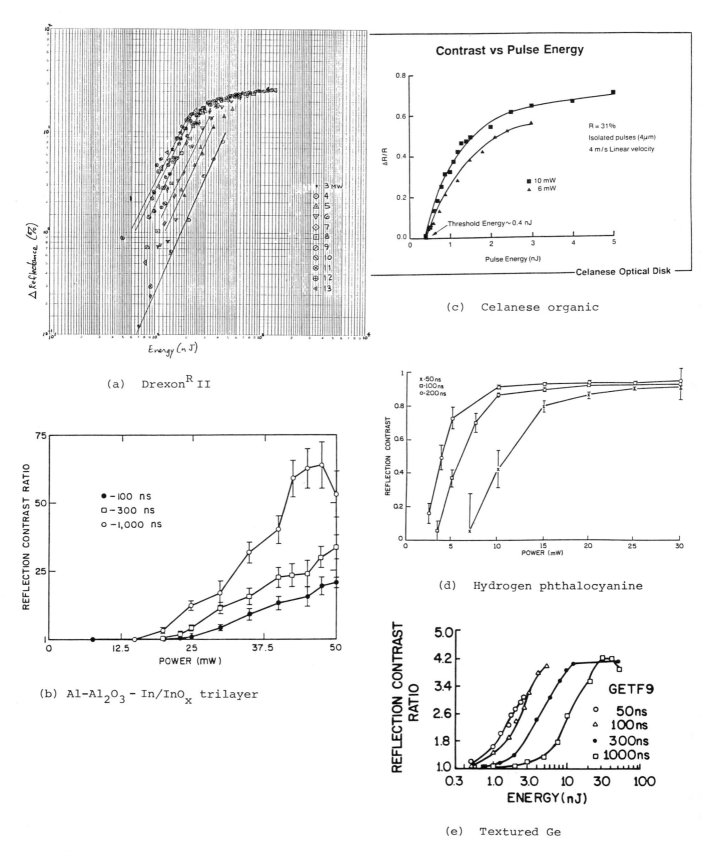

(a) Drexon[R] II

(b) Al–Al$_2$O$_3$ – In/InO$_x$ trilayer

(c) Celanese organic

(d) Hydrogen phthalocyanine

(e) Textured Ge

Figure 3.   Measured sensitivity functions for five types of WORM media

In the following, we will apply this generalized expression to specific cases of WORM media.

Case 1°. __Drexon II WORM media.__ The applicable references [3,4] have given a very complete description of the Drexon II media and set of experiments. Figure 4 shows typical 1/E plots of the data. The experimental evidence is that E is the dominant independent variable for Drexon II and that P is the secondary variable [3]. Also, there is a single lim ΔR which is independent of E or P (at large E or P), and X_T^{α} ($=E_T^{\alpha}$) is a function of incident laser power P, see Fig. 5. A closed form expression for the sensitivity then is readily derivable:

$$\Delta R(E,P) = \Delta R(\infty)[1-E_T^{\alpha}(P)/E] \qquad (3)$$
$$\Delta R(\infty) = 29\% \qquad (4)$$
$$1/E_T^{\alpha}(P) = [1/E_T^{\alpha}(\infty)][1-P_T^{\alpha}/P] \qquad (5)$$
$$E_T^{\alpha}(\infty) = 80 \text{ nJ} \qquad (6)$$
$$P_T^{\alpha} = 3 \text{ mW.} \qquad (7)$$

Figure 6 shows the agreement between theory and experiment over the full range of measured E and P.

Case 2°. __Al-Al$_2$O$_3$-In/InO$_x$ Trilayer.__
Figure 7 shows a 1/P plot, indicating that P, not E is, the dominant independent variable for this WORM media [5]. The threshold is a constant, but the saturation ΔR is a function of laser pulse length t. Thus, the general sensitivity function, following our prescription, is

$$\Delta R(P,t) = \lim \Delta R(t)[1-P_T^{\alpha}/P] \qquad (8)$$
$$P_T^{\alpha} = 27 \text{ mW} \qquad (9)$$
$$\lim \Delta R(t) = 37 + 0.12t, \quad t \text{ in nsec.} \qquad (10)$$

The data for 1000 nsec pulses at high incident laser powers indicate severe saturation or material modification. Therefore, these extreme data were deleted in the derivation of lim $\Delta R(t)$. The derived expression suggests that the extreme saturation effects have an onset at pulse lengths of about 500 nsec.

Case 3° __Celanese Organic Media.__ Celanese, quite properly, have measured their media with respect to incident laser energy [6,7]. Figure 8 shows the 1/E plot for their media. We find that the fit is very good, but that there is a residual dependence on incident laser power for the threshold and the saturation level. The 6 mW and 10 mW 1/E plots, which are essentially parallel, suggest an expression of the form

$$\Delta R(E,P) = [\Delta R(\infty)(1-\tilde{P}_T^{\alpha}/P)][1-E_T^{\alpha}(1-P_T^{\alpha}/P)/E] \qquad (11)$$

which can be unfolded because

$$E_T^{\alpha}(1-P_T^{\alpha}/P) = \begin{cases} 0.645 @ P = 6 \text{ mW} & (12a) \\ 0.57 @ P = 10 \text{ mW} & (12b) \end{cases}$$

$$\Delta R(\infty)(1-\tilde{P}_T^{\alpha}/P) = \begin{cases} 0.7 @ P = 6 \text{ mW} & (13a) \\ 0.8 @ P = 10 \text{ mW} & (13b) \end{cases}$$

This media model has four constants which are experimentally well defined from the four x and y axis intercepts. We obtain from these intercepts

$$\Delta R(\infty) = 0.95 \qquad (14)$$
$$\tilde{P}_T^{\alpha} = 1.58 \qquad (15)$$
$$E_T^{\alpha} = 0.46 \qquad (16)$$
$$P_T^{\alpha} = -2.46 \qquad (17)$$

and, thus,

$$\Delta R(E,P) = [0.95 (1-1.58/P)][1-0.46(1+2.46/P)/E], \qquad (18)$$

This equation explicitly gives the incident power dependence of the saturation and threshold levels. More power series tests need to be taken on this media to explain the observed behavior.

Case 4°. __Hydrogen Phthalocyanine Thin Films.__ Figure 9 shows that, when the abscissa of the reflection contrast graphs is converted from incident power to incident energy, that the resultant graphs are essentially overlayed [8]. This shows that the hydrogen phthalocyanine behavior has incident energy, rather than power, as its primary independent variable. If we now plot 1/E graphs for the 50, 100 and 200 nsec data, then we have Figure 10, which shows a behavior very similar to the Drexon II. A plausible expression for this WORM media is given

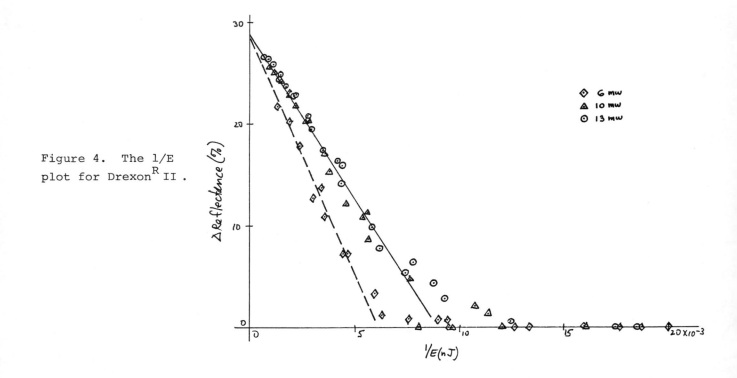

Figure 4. The 1/E plot for Drexon^R II.

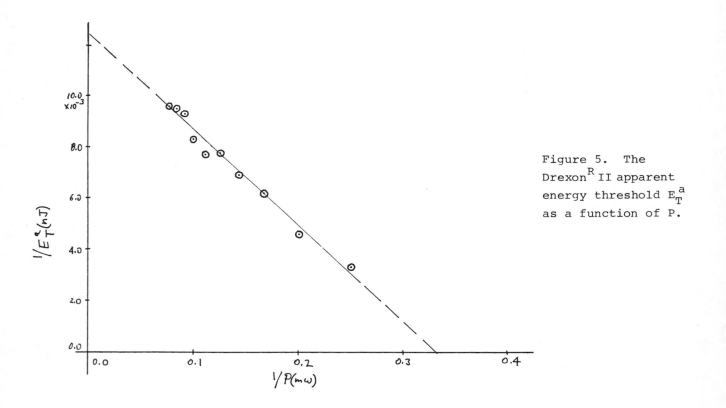

Figure 5. The Drexon^R II apparent energy threshold E_T^a as a function of P.

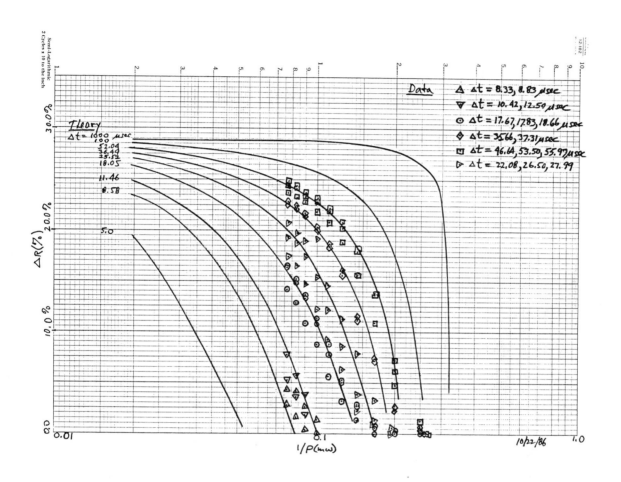

Figure 6. Comparison of experimental and theoretical 1/P plots for Drexon[R] II.

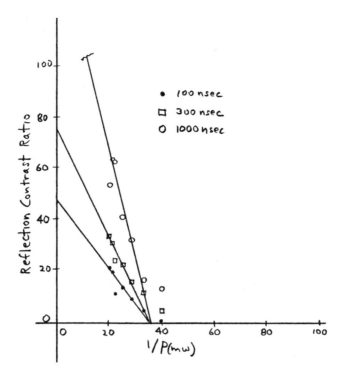

Figure 7.   The 1/P plot for Al-Al$_2$O$_3$-In/InO$_x$ trilayer.

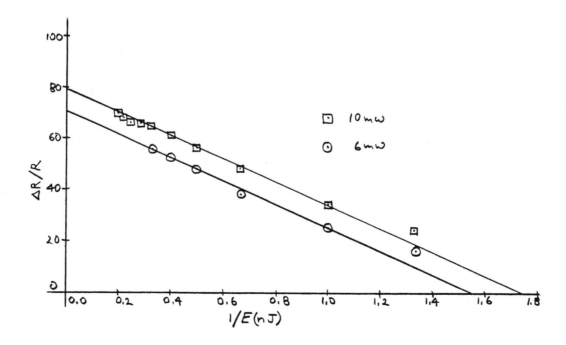

Figure 8.   The 1/E plot for Celanese organic media.

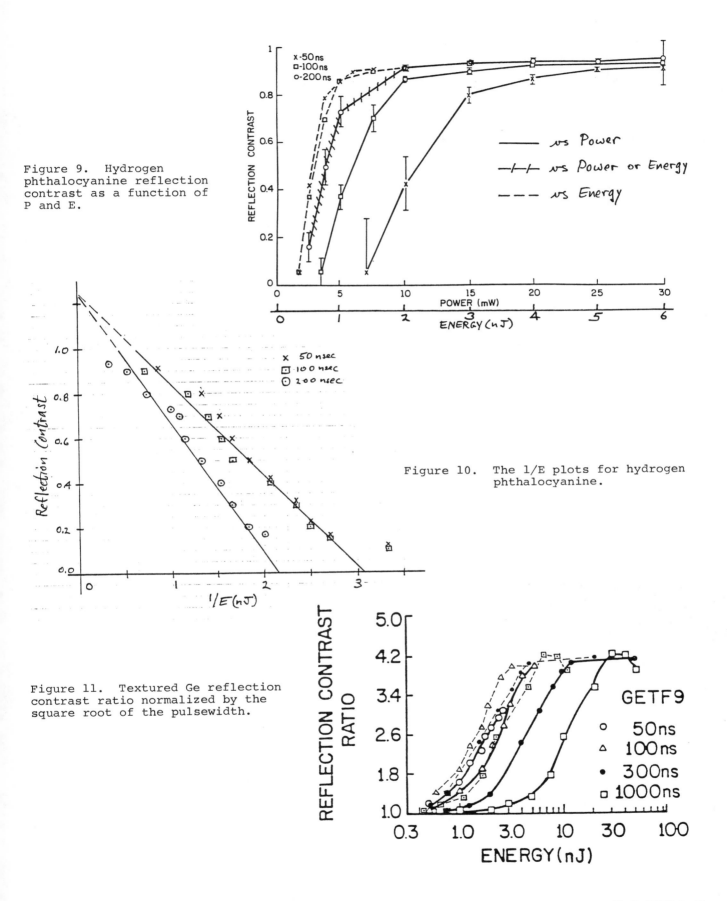

Figure 9. Hydrogen phthalocyanine reflection contrast as a function of P and E.

Figure 10. The 1/E plots for hydrogen phthalocyanine.

Figure 11. Textured Ge reflection contrast ratio normalized by the square root of the pulsewidth.

by

$$\Delta R(E,t) = \Delta R(\infty)[1-E_T^a(t)/E] \qquad (19)$$
$$\Delta R(\infty) = 125\% \qquad (20)$$
$$E_T^a(t) = E_T^a(1-\tau/t) \qquad (21)$$
$$E_T^a = 0.61 \text{ nJ} \qquad (22)$$
$$\tau = 45 \text{ nsec.} \qquad (23)$$

This model is good for pulses up to about 1.3 nJ, beyond which the media saturates (depending on pulse width). Note that the data and the model suggests that hydrogen phthalocyanine becomes resistive to laser marking at the longer pulse lengths - the apparent energy threshold for laser marking increases with pulse width. The saturation of ΔR at pulse energies around 1nJ suggests a media model which is a hyperbolic function of 1/E, rather than a linear function. We need more data to clarify this issue.

Case 5°. Textured Media. Figure 11 shows the same data as Figure 3d, but normalized by the square root of the pulsewidth, $(t_o/t)^{1/2}$ where t_o = 50 nsec. The normalized curves (broken lines) are essentially overlayed. It appears that E is the dominant and t the secondary independent variable. The square root dependence on t suggests a strong thermal diffusion effect. Also, the media appears to resist laser marking at the longer pulse lengths. Figure 12 then shows a 1/P plot for the reflection contrast data of 4 minute etched samples measured with 100 nsec pulses [9]. A straight line fit to the 1/E plot gives

$$\Delta R(E) = \Delta R(\infty)[1-E_T^a/E], \qquad (24)$$
$$\Delta R(\infty) = 146\%, \qquad (25)$$
$$E_T^a = 0.95 \text{ nJ}. \qquad (26)$$

A plausible and general sensitivity function for this media is given by

$$\Delta R(E,t) = \Delta R(\infty)[1-(t_o/t)^{1/2} E_T^a/E] \qquad (27)$$

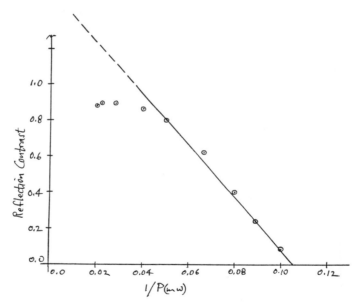

Figure 12. The 1/P plot for textured Ge

The textured media behaves somewhat like hydrogen phthalocyanine, especially with respect to the pulsewidth dependency. A saturation of the sensitivity function commences at about 25 mW, so the above expression is valid for incident laser powers less than this value. Just as in the case of hydrogen phthalocyanine, it is tempting to model this media as a hyperbolic, rather than linear, function of 1/E, with pulse width as a discriminant. However, there aren't enough data at this time to pursue this line of analysis.

5. CONCLUDING REMARKS

It has been shown that a general theoretical formalism exists which can be used to describe a wide variety of WORM media. This formalism can be used to analyze empirical sensitivity data. The proper methodology requires that the experimenter include pulse length (or energy or energy density), as well as power series, in setting up a test plan for evaluating a particular developmental WORM media.

The new formalism is powerful because it explicitly identifies the experimental write laser parameters which affect the threshold and saturation phenomena. Thus, this formalism can be used to guide a detailed physical analysis of the laser marking phenomenon for a media.

Finally, the new formalism specifically identifies the pulse width dependency of a media, which would show up as a spurious amplitude modulation in a typical pulse code modulation scheme. The closed form expressions derived from the new general formalism permits a better analysis of signal to noise for a given media at a given wavelength and a given modulation scheme.

7. ACKNOWLEDGMENTS

The author gratefully acknowledges the encouragement and support of Prof. E. Garmire, Director of the Center for Laser Studies. Technical discussions with J. Kaldem of Drexler Technology Corporation and Dr. S. Y. Suh of the Hoechst Celanese Corporation provided helpful insight into the behavior of WORM media, and these are gratefully acknowledged. The paper was supported by California Peripherals, Inc., and Mendez R&D Associates, and this support also is gratefully acknowledged.

8. REFERENCES

[1] E. V. LaBudde, R. A. LaBudde, and C. M. Shevlin, "Theoretical Modelling, Calculations, and Experiments Characterizing the Laser Induced Hole Formation Mechanism of an In-Contact Overcoated Optical Disk Medium", Proc. SPIE 382, 116 (1983).

[2] E. V. LaBudde and R. A. LaBudde, "A Generalized Mathematical Model of Noise Sources Affecting Optical Recording Processes", Topical Meeting on Optical Data Storage, Optical Society of America, October, 1985.

[3] A. J. Mendez, "Application of a Generalized Mathematical Model of Noise Sources to Optical Media Evaluation", Proc. SPIE 695, 373 (1986).

[4] R. A. LaBudde, E. V. LaBudde, and R. M. Hazel, "Theoretical Modelling and Experimental Characterization of Drexon^R Optical Media", Proc. SPIE 382, (1983).

[5] A. F. Hebard, G. E. Blonder, and S. Y. Suh, "Optical Recording Applications of Reactive Ion Beam Sputter Deposited Thin Film Composites", Appl. Phys. Lett 44 1023 (1984).

[6] H. A. Goldberg, R. S. Jones, P. S. Kalyanaraman. R. S. Kohn, J.E. Kuder, and D. E. Nikles, "Dynamic Laser Marking Experiments on an Organic Optical Storage Medium", Proc. SPIE 695, 20 (1986).

[7] J. E. Kuder, "Organic Active Layer Materials for Optical Recording", Proc. 40th Annual Conference SPSE, Rochester, NY, May 1987.

[8] M. F. Dautartas, S. Y. Suh, S. R. Forrest, M. L. Kaplan, A. J. Lovinger, and P. H. Schmitt, "Optical Recording Using Hydrogen Phthalocyanine Thin Films", Appl. Phys. (A) 36, 71 (1985).

[9] S. Y. Suh, D. A. Snyder, D. L. Anderson, and H. G. Craighead, "Dynamic Signal to Noise Ratio Measurements of Textured Optical Storage Media", Appl. Optics 23, 3956 (1984).

[10] S. Y. Suh, "Optical Writing Process in Textured Media", Proc. SPIE 382, 196 (1983).

[11] J. J. Wrobel, "The Physics of Recording in Write Once Optical Storage Materials", Proc. SPIE 420, 288 (1983).

[12] D. G. Howe and A. B. Marchant, "Digital Optical Recording in Infrared Sensitive Organic Polymers", Proc. SPIE 382, 103 (1983).

Fast nematic liquid crystal spatial light modulator

D. Armitage, J. I. Thackara, W. D. Eades,
M. A. Stiller, W. W. Anderson

Lockheed Missiles & Space Company, Inc., Research & Development Division
3251 Hanover Street, Palo Alto, California 94304-1187

ABSTRACT

Electro-optic shutters that operate at a kilohertz rate have been demonstrated with nematic liquid crystals operated in the surface-mode or π-cell configuration. This principle has been employed to fabricate an NLC photoaddressed spatial light modulator that is faster than twisted-nematic devices.

1. INTRODUCTION

Nematic liquid crystal (NLC) electro-optic technology satisfies the requirement for low-voltage, low-power, inexpensive optical modulators, provided the speed requirement is modest. The sensitive electro-optic behavior of the NLC results from a fluidity that allows reorientation of the NLC optic axis. Viscous retardation is an unwanted consequence of this motion and restricts the response speed. NLC materials development has reached such a sophisticated level that significant further reduction of the viscosity is improbable. Therefore, the only way to increase the response speed is to increase the driving force.

The reorientation torque induced in an NLC by an applied electric field arises from an induced dipole moment; therefore, the sign of the torque is independent of the sign of the electric field. Conventional NLC devices developed by the display industry employ twisted-nematic structures that have a driven on-state, but rely on the NLC elasticity in returning to the off-state.[1] The elastic restoring force can only be increased by increasing the distortion of the NLC; however, excessive distortion is incompatible with the twisted-nematic geometry. Variable or tunable birefringence NLC devices can be operated with a strongly distorted NLC configuration that favors fast response.

Birefringent optical modulation in the 10-kHz range has been demonstrated with a NLC device. The birefringence was restricted to a region close to the electrode surfaces, hence the expression "surface-mode device."[2,3] The π-cell is a similar device with improved viewing angle for display applications.[4] Twisted-nematic structures provide threshold devices that facilitate matrix-addressing and are not sensitive to voltage and NLC cell tolerances. The surface-mode device sacrifices the threshold effect for response speed and consequently is sensitive to fabrication nonuniformities.

Two other methods of increasing the frame speed of NLC devices have been reported. The first employs NLC with a dielectric anisotropy that changes sign with frequency, allowing drive-on and drive-off in a two-frequency addressing scheme.[1] The second method varies the orientation of the drive electric field in a transverse electric field configuration.[5]

We have fabricated a photoaddressed spatial light modulator (SLM) employing the surface-mode NLC configuration. Single-crystal silicon provides the photoaddressing medium in order to avoid speed limitations imposed by photoconductor detrapping times. Replication optics is employed in the retroreflective SLM readout mirror to provide the necessary optical quality. Preliminary results indicate the following characteristics: write time ~100 μs, erase time ~100 μs, frame rate ~1 kHz, 50 percent modulation transfer function (MTF) ~ 5 lp/mm, limiting MTF ~20 lp/mm, sensitivity ~10 nJ/cm^2, optical quality ~0.5 wave over 25-mm aperture, and a peak diffraction efficiency approaching the theoretical limit of 34 percent.

2. NLC DEVICE THEORY

Figure 1 illustrates a liquid-crystal cell with uniform parallel alignment and thickness L. Transparent electrodes provide an electric field perpendicular to the surface while allowing optical detection of the nematic response. The electric field (E) gives rise to a bulk torque due to the dielectric anisotropy ($\Delta\epsilon$).

$$\text{Electric Torque} = E^2 \epsilon_o \Delta\epsilon \sin\phi \cos\phi \qquad (1)$$

An elastic torque due to NLC optic axis or director ($\hat{n}$) spatial distortion can be written in terms of an average elastic constant (K).

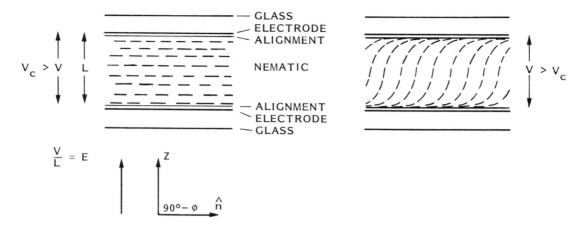

Figure 1. Liquid-crystal cell.

$$\text{Elastic Torque} = K\partial^2\hat{n}/\partial z^2 \tag{2}$$

In static equilibrium the torques balance and an expression for the spatial dependence of $\hat{n}$ can be derived. It can be shown that there is no response until a critical voltage V_c is exceeded. The critical voltage is independent of cell thickness since L influences the elastic term via the boundary conditions where strong anchoring is assumed.

$$V_c = \pi(K/\epsilon_o \Delta\epsilon)^{1/2} \tag{3}$$

Substantial director reorientation corresponds to voltages slightly higher than critical. The form of the director field is illustrated in Fig. 1. Behavior of this general nature is called a Frederiks transition and can be induced by electric, magnetic, or flow fields.[1]

The dynamics of the Frederiks transition are complex with higher spatial gradients of the director field responding at faster rates. A crude approximation with some validity in device work can be written in terms of an average viscosity (η).[1]

$$\tau_r = \eta L^2/(\epsilon_o \Delta\epsilon V^2 - K\pi^2) \tag{4}$$

$$\tau_d = \eta L^2/(K\pi^2) \tag{5}$$

where τ_r and τ_d are exponential time constants describing the rise and decay rates. The expression for τ_r greatly underestimates the response time when the nematic director and electric field approach an orthogonal or parallel alignment. This follows from the vanishing of the electric torque term under these conditions.

If the applied voltage is maintained above critical (V_c) and varied to modulate the cell retardation, a much faster response results. Under these conditions the angle ϕ in Eq. (1) tends towards 45° over the active cell region, thus maximizing the torque term. Figure 2 shows a cell operated well above critical voltage. In the bulk of the cell, the nematic director is aligned along the optical propagation direction. Birefringence is restricted to a thin surface layer at each boundary surface, hence the description surface-mode cell.[2,3] The cell can be assembled with complementary tilt or opposing tilt bias and is called a π-cell in the latter case. The π-cell compensates somewhat for off-axis viewing in display applications.[4]

A detailed analysis of the structure shown in Fig. 1 yields the expression:[6]

$$\left(\frac{d\phi}{dz}\right)^2 = \frac{D_z^2\gamma(\sin^2\phi_m - \sin^2\phi)}{\epsilon_o\Delta\epsilon K_{11}(1 + k\sin^2\phi)(1 + \gamma\sin^2\phi)(1 + \gamma\sin^2\phi_m)} \tag{6}$$

PARALLEL SURFACE ALIGNMENT

TILT COMPLEMENTARY

TILT OPPOSED
π MODE

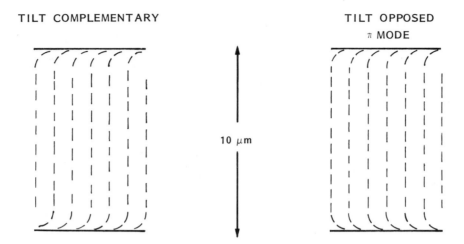

10 μm

BIREFRINGENCE RESTRICTED TO SURFACE REGION $< 1\ \mu m$

Figure 2.   Surface-mode liquid-crystal cells.

where $D_z = \epsilon_o E_z(\epsilon_\perp \cos^2\phi + \epsilon_{||} \sin^2\phi)$; applied volts $= \int_0^L E_z dz$; $\phi \to 0$ when $z \to 0$ or L, and defines rigid boundary conditions; $\gamma = \Delta\epsilon/\epsilon_\perp$; K_{11} = splay elastic constant, K_{33} = bend elastic constant, and $k = (K_{33} - K_{11})/K_{11}$; ϕ_m = maximum director tilt angle; and E_z = electric field.   Considerable simplification is achieved in the high-voltage approximation: $V \gg V_c = \pi(K_{11}/\epsilon_o\Delta\epsilon)^{1/2}$, when $\phi_m \to \pi/2$.

In addition, assuming $k = \gamma$ reduces Eq. (6) to the form:

$$\frac{d\phi}{dz} = \frac{\pi D_z \cos\phi}{V_c \epsilon_o (\epsilon_\perp \epsilon_{||})^{1/2}(1 + \gamma \sin^2\phi)} \qquad (7)$$

The condition $k = \gamma$ is a reasonable approximation for many nematics of interest; e.g., for E7, $k = 0.93$ and $\gamma = 2.56$.   It can be shown that in this case, substitution of an effective $\gamma' = (\gamma + k)/2$ leads to errors of less than 5 percent in Eq. (7).

The applied voltage V is related to D_z via the expression:

$$\frac{V}{2} = \int_0^{L/2} E_z dz = \frac{D_z}{\epsilon_o} \int_0^{L/2} \frac{dz}{\epsilon_\perp \cos^2\phi + \epsilon_{||} \sin^2\phi} \qquad (8)$$

where $\phi = \phi_m = \pi/2$ at $z = L/2$.

Eq. (8) can be written:

$$V = \frac{2D_z}{\epsilon_o \epsilon_\perp} \int_0^{L/2} \frac{dz}{1 + \gamma \sin^2\phi} \qquad (9)$$

Substituting Eq. (7) in Eq. (9) and integrating,

$$D_z = \frac{\epsilon_0 \epsilon_{\parallel}}{L} \left(V - \frac{2V_c}{\pi (1 + \gamma)^{1/2}} \right) \tag{10}$$

when $V \gg V_c$, $D_z \rightarrow V \epsilon_0 \epsilon_{\parallel} / L$.

The birefringent retardation δ waves per surface can be approximated:

$$\delta = 1/\lambda \int_0^{L/2} (\Delta n - 0.75 \, \Delta n^2/n_0) \cos^2 \phi \, dz \tag{11}$$

Substituting Eqs. (7) and (10) in Eq. (11) and integrating in the limit $V \gg V_c$

$$\delta = \frac{V_c L (\Delta n - 0.75 \, \Delta n^2/n_0)(1 + \gamma'/3)}{\pi V \lambda} (\epsilon_{\perp}/\epsilon_{\parallel})^{1/2} \tag{12}$$

where $\gamma' = (k + \gamma)/2$.

A more accurate theory involving numerical integration is available.[7]

In the high field limit necessary for fast response, the retardation is proportional to L/V. Therefore, given a range of retardation, the ratio L/V is fixed, hence the charge required is essentially a constant independent of the particular L or V employed. The cell capacitance per unit area approaches

$$C = \frac{\epsilon_{\parallel} \epsilon_0}{L} \tag{13}$$

From Eq. (12) changes in retardation can be written

$$\frac{d\delta}{dV} = \frac{-V_c L (\Delta n - 0.75 \, \Delta n^2/n_0)(1 + \gamma'/3)(\epsilon_{\perp}/\epsilon_{\parallel})^{1/2}}{(V^2 \pi \lambda)} \tag{14}$$

The associated charge density Q is determined by the expression

$$C = \frac{dQ}{dV} \tag{15}$$

Substituting Eqs. (13) and (15) in Eq. (14) gives the charge sensitivity:

$$\frac{d\delta}{dQ} = \frac{-V_c L^2 (\Delta n - 0.75 \, \Delta n^2/n_0)(1 + \gamma'/3)\epsilon_{\perp}^{1/2}}{V^2 \pi \lambda \epsilon_{\parallel}^{3/2} \epsilon_0} \tag{16}$$

The dynamics of the Frederiks transition has been analyzed in detail for magnetic field deformations.[8] The analysis applies here, with the adoption of electric rather than magnetic free-energy formulations. Translating the magnetic result into electrical terms in the high field regime gives:

$$\tau_V = \frac{\eta L^2 V_c^2}{K_{33} \pi^2 V^2} \tag{17}$$

where τ_V is the response at voltage V.

Substituting Eq. (17) in Eq. (16)

$$\frac{d\delta}{dQ} = \frac{\tau_V K_{33}}{\eta \lambda} \left(\frac{\epsilon_{\perp} \Delta \epsilon}{\epsilon_{\parallel}^3 \epsilon_0 K_{11}} \right)^{1/2} \left[\left(\Delta n - 0.75 \, \Delta n^2/n_0 \right) (1 + \gamma'/3) \right] \tag{18}$$

Substituting the values for nematic eutectic mixture E7 (EM Chemicals):

$$K_{11} = 10.7 \times 10^{-12} \text{ N}$$
$$K_{33} = 20.7 \times 10^{-12} \text{ N}$$
$$\epsilon_{\parallel} = 19.2$$
$$\epsilon_{\perp} = 5.4$$
$$\eta = 0.38 \text{ poise}$$
$$\Delta n = 0.224$$
$$n_o = 1.5222$$
$$\gamma' = 1.75$$

and response time $\tau_V = 400 \ \mu s$ with $\lambda = 0.516 \ \mu m$, gives:

$$\frac{d\delta}{dQ} = 1.41 \times 10^3 \text{ waves-m}^2/\text{C per surface.}$$

In a reflective NLC SLM, four surface regions are involved and hence the total phase peak-to-peak modulation (2Γ) can be written in terms of the charge modulation Q_p (C/m^2)

$$2\Gamma = 5.65 \times 10^3 \ Q_p \text{ waves.} \tag{19}$$

3. PHOTORECEPTOR

In a photoaddressed SLM the photoreceptor converts photons into charge that then acts on an electro-optic element to provide the desired output optical modulation. Figure 3 illustrates the SLM structure. The quantum efficiency, spatial resolution, response speed, and influence on optical quality are characteristics of the photoreceptor. Polycrystalline photoconductors are limited in response speed by detrapping times. Therefore, making speed a priority implies a monocrystal photoreceptor, and silicon technology is obviously favored. A silicon wafer with photodiode input and a resolution-preserving microdiode array serves as an effective photoreceptor.[9] However, the p-n junction technology employed is sensitive to contamination and electrostatic effects and requires diffusion furnace temperatures. These factors make it difficult to achieve optical wavefront quality in the fabrication and assembly of the SLM.

An effective NLC SLM photoaddressed via a bare silicon or gallium arsenide monocrystal wafer has been reported.[10] We have investigated this approach, which can be described in semiconductor device terminology as a metal-insulator-semiconductor-insulator-metal structure, where the NLC forms the latter insulator. However, we prefer the term photovaractor for this and similar structures. In this device, charge is photo-excited and transported as in the photodiode, while low surface mobility is required to preserve resolution. The charge is confined to the silicon, hence electron-hole recombination is the erasure mechanism.

High-resistivity silicon of kilohm-centimeter order with consequent low depletion voltage is essential to both the photodiode and photovaractor addressing mechanisms. The leakage or dark current in a photoreceptor of such high bulk resistivity is dominated by surface mechanisms. Surface contamination associated with standard semiconductor cleaning processes can result in high leakage and an ineffective photoreceptor.[11] The optical cements used in assembly generate large electrostatic fields that charge the silicon surfaces and result in excessive leakage. A simple solution to these problems is to coat the silicon surface with a thin film of high-resistance amorphous silicon-hydrogen alloy (aSi:H). The surface contamination and charging effects are now confined to the aSi:H layer, where low carrier mobility suppresses the leakage current. The aSi:H layer is thin; consequently, detrapping leakage current effects are not significant. Figure 3 shows the resultant aSi:H/silicon wafer photoreceptor in assembled form.

The SLM resolution is degraded by lateral charge drift in the aSi:H region and charge defocussing in the depletion region.[12] These effects can be reduced by etching a grid in the silicon, as shown in Fig. 4, to provide dielectric barriers.[13] The mirror replication process fills the etched grid with insulating cement.

4. TEST

The experimental results are for SLMs fabricated from smooth ungrooved silicon photoreceptors and replicated dielectric mirrors.

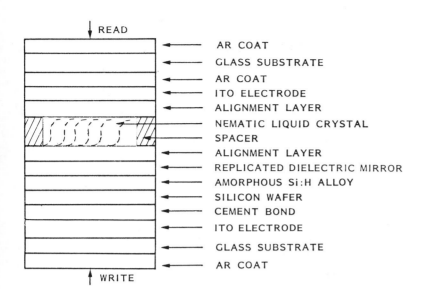

Figure 3. Liquid-crystal SLM structure.

Figure 4. Charge-confining dielectric barriers.

The NLC SLM is driven by a voltage pulse coupled through a capacitor-resistance filter to remove the dc component, thereby minimizing electrochemical effects in the NLC. Figure 5 shows the integrated dark current or charge response of the device to a 60-V, 1-ms pulse. The depletion voltage for the silicon is 19 V. The depletion charge plus capacitive charge equals 100 nC for the 20-cm^2 area, at 60 V applied. The experimentally recorded charge is 200 nC for the polarity aSi:H negative. In the reverse polarity a larger charge flow is recorded due to the negative charge accumulated at the bare silicon-cement interface. This asymmetry is similar to diode behavior, but here the charge is confined to the silicon. The dark charge biases the NLC in the surface mode configuration.

Figure 6 shows the general purpose optical test bed. An argon laser provides the optical write and read pulses at 514 nm wavelength. A helium-neon laser is included in the setup for other experiments not discussed here. Timing and pulse length control are achieved by slaving two pulse generators to a master signal generator. Acousto-optic modulators control the optical write and read pulses. The Twyman-Green interferometer in the write path provides a grating input for MTF measurements. Alternatively, a test bar chart can be imaged onto the write side. The TV cameras in the readout image and focal planes facilitate output analysis. A photodiode detector in the focal plane is included for quantitative MTF and response speed measurements. An additional Twyman-Green

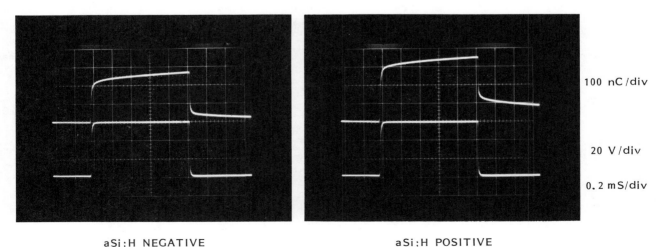

Figure 5. SLM voltage-charge characteristic: lower trace 20 V/div, upper trace 100 nC/div, 0.2 ms/div, 100 Hz frame rate.

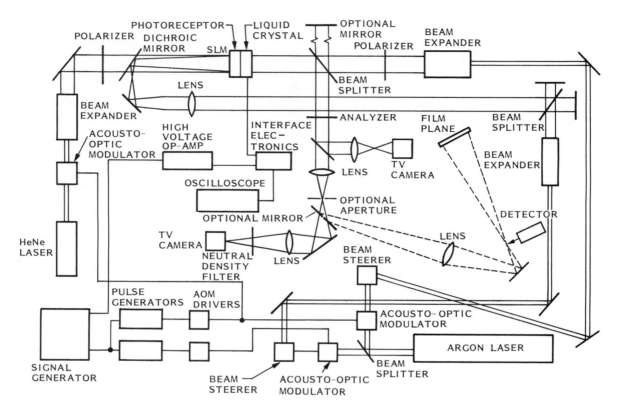

Figure 6. SLM test bed.

interferometer in the output path provides a phase map that extends the MTF data to low spatial frequencies and allows the output optical quality to be assessed under dynamic operating conditions.

Figure 7 illustrates the behavior of the NLC SLM as a phase modulator. The NLC cell is parallel surface aligned and untwisted, with the nematic director along the direction of read light polarization to maximize the phase modulation. The read interferogram is recorded under operating conditions in the absence of write light. When the write interferogram modulates the read interferogram, the product interferogram results. The phase modulation can be measured directly, as indicated in Fig. 7, and is expressed as a sinusoidal amplitude at spatial frequency w in the x-direction:

$$\text{Phase modulation} = \Gamma \sin (2 \pi w x) \tag{19}$$

The output pictures reveal nonuniformities and defects present in the device at this stage of development. The dynamic optical quality achieved so far is double pass 0.7 wave peak-to-valley error at 514 nm wavelength over a 1-in. aperture.

Dispensing with the read interferometer and writing an interferometric input provides a phase grating of amplitude Γ having a diffraction efficiency (DE) determined by the first order Bessel function (J_1):

$$\text{Diffraction efficiency} = J_1^{2} (\Gamma) \tag{20}$$

The MTF is quantified by recording DE and Γ as a function of w. Figure 8 shows the result, where the DE data are more reliable at high spatial frequency and conversely for the direct Γ data.

Figure 9 shows the operating conditions for the MTF data and time response of the first-order diffraction spot. The first order diffracted light is measured by a photodiode, which allows the time evolution to be displayed. A DE of 10 percent is achieved in a writing time of 300 μs. The diffraction intensity decays with a time constant of approximately 1 ms on cessation of the write pulse. The NLC grating can be erased by a negative pulse, as shown in Fig. 9, which provides a uniform erasure charge that exceeds the NLC dynamic range. The repetition rate is maintained at 100 Hz throughout the experiments.

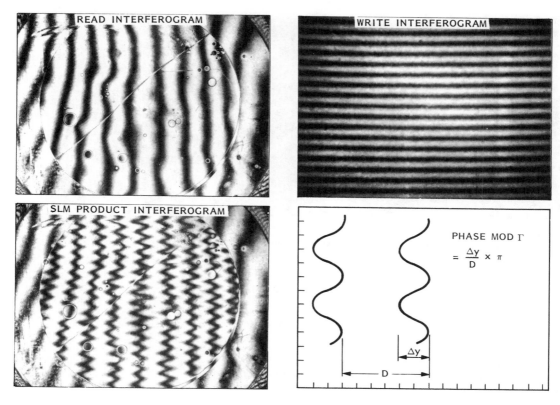

Figure 7. Low-frequency phase modulation test.

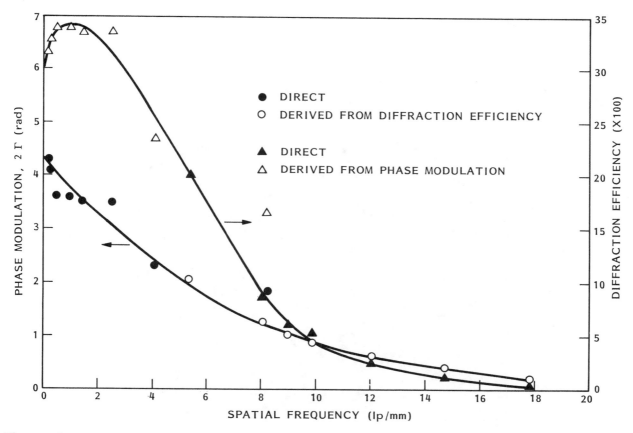

Figure 8. SLM diffraction efficiency and phase modulation against spatial frequency.

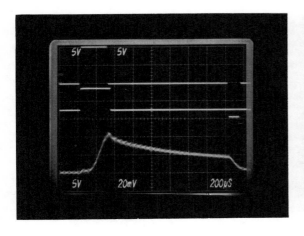 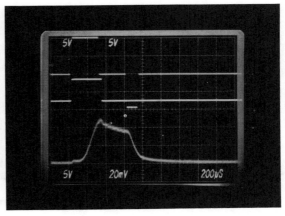

SLM 50 V/div

WRITE LIGHT

DIFFRACTION
INTENSITY

Figure 9.   Response speed of first order diffraction intensity at 100 Hz frame rate.

The decay rate is slower than predicted by NLC theory, and it would not be surprising to find the simple exponential approximation to be inadequate. We suspect that the SLM can be overwritten rather than erased, but have not yet performed the experiment. The overwrite mode will provide the maximum frame rate, which we anticipate to be in the 1-kHz region. We have demonstrated frame rates of 700 Hz in the write-erase mode.

A uniform write-light illumination over an area of 4.9 cm^2 produces a measured photocharge of 37.5 $\pm$ 2.5 nC when operated with 100-V, 400-μs pulses at 100-Hz and 400-μs write-light pulses. The write pulse optical energy was derived from a measured continuous power of 0.068 mW/cm^2 multiplied by the 400-μs pulse length. For the photon energy 2.41 eV, the quantum efficiency is calculated at 67.5 $\pm$ 4.5 percent. The reflectivity loss was measured as 20 percent; therefore, ignoring the 1 percent ITO absorption, the quantum efficiency could be raised to 84.5 $\pm$ 5.5 percent by antireflective-coating the silicon. A DE of 5.7 percent was measured at 12 lp/mm with write-light intensity 0.068 mW/cm^2 and pulse length 250 μs, i.e., 17-nJ pulse energy. The peak charge corresponding to this sinusoidal light input is 9.5 nC/cm^2 and the DE corresponds to a peak-to-peak modulation of 1 rad. This is comparable with the theoretical prediction.

The bar-chart image in Fig. 10 shows the uniformity achieved at this early stage of development. In this test the read polarization is turned at 45° to the NLC axis to provide birefringent amplitude modulation in a crossed polarizer configuration. The discernible resolution approaches 10 lp/mm. The SLM is operated on the last dark cycle of birefringence for best uniformity and speed.

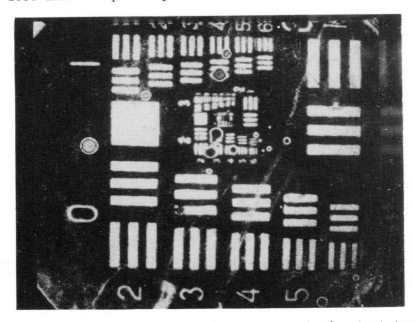

Figure 10.   SLM image of U.S. Air Force resolution test target.

5. CONCLUSION AND FUTURE DIRECTIONS

A photoaddressed NLC SLM that pushes the NLC response speed to the attainable limit has been demonstrated. The fabrication methods are consistent with high optical quality and uniformity. The resolution demonstrated at this stage falls well short of the resolution intrinsic to a thin NLC layer.

The replicated mirror results in some loss of resolution via field fringing, but does not account for the observed resolution limit. The inherent variation in cement thickness influences the uniformity of voltage transfer to the NLC. However, provided leakage current is low, the voltage levels are set by charge transfer constraints. This means that thicker regions of cement (or NLC) are provided with more voltage at the expense of the silicon voltage. The surface figure of the silicon wafer could be improved by an assembly technique that forces the wafer to conform to an optically flat surface during the cementing process.

An alternative to mirror replication is to polish the mounted silicon wafer. This should allow the surface to be polished to an adequate optical figure while preserving the electronic surface quality.[16] A semi-insulating film such as aSi:H could then be deposited without an excessive rise in temperature. The dielectric mirror could then be directly deposited. Thickness variations in the cement bonding the silicon to the substrate are not important because that silicon surface is highly conducting. This fabrication method should be a better method to promote uniformity and, by further thinning the silicon, could result in higher resolution.

An amorphous-crystal silicon interface acts as a rectifying heterojunction.[17] The influence of this junction on device performance has not been analyzed at this stage.

Optical processing demands still higher speed than can be provided by NLC devices. This demand is expected to be fulfilled in the near future by ferroelectric liquid crystal (FLC) devices. We also are developing a FLC SLM employing silicon photoaddressing technology.[18]

6. ACKNOWLEDGEMENTS

Technical assistance from S. K. Ichiki, C. L. Valencia, and E. N. Okazaki, and technical discussion with the late J. H. Becker, are acknowledged. This work was supported by Rome Air Development Center and currently is supported by a Lockheed Missiles & Space Company, Inc., Research & Development Division program.

7. REFERENCES

1. E. B. Priestley, P. J. Wojtowicz, and P. Sheng, eds., Introduction to Liquid Crystals, Plenum, New York (1974).

2. J. L. Fergason, "Performance of matrix display using surface mode," Conference Record, Biennial Display Research, 177-182 (1980).

3. J. L. Fergason, "Use of strong surface alignment in nematic liquid crystals for high speed light modulation," in Liquid Crystals and Spatial Light Modulator Materials, William A. Penn, ed., Proc. SPIE 684, 81-86 (1986).

4. P. J. Bos, P. A. Johnson, and K. R. Koehler/Beran, "The Π-cell: A new, fast liquid-crystal optical switching device," Mol. Cryst. Liq. Cryst. 113, 329-337 (1985).

5. D. Armitage and J. I. Thackara, "Liquid-crystal differentiating spatial light modulator," in Nonlinear Optics and Applications, Pochi Yeh, ed., Proc. SPIE 613, 165-171 (1986).

6. H. J. Deuling, "Elasticity of nematic liquid crystals," in Solid State Physics, Supplement 14, L. Liebert, ed., Academic Press (1978).

7. D. A. Balzarini, D. A. Dunmur, and P. Palffy-Muhory, "High voltage birefringence measurments of elastic constants," Mol. Cryst. Liq. Cryst. 102 (Letters), 35-41 (1984).

8. P. Pieranski, F. Brochard, and E. Guyon, "Static and dynamic behavior of a nematic liquid crystal in a magnetic field, Part 2: Dynamics," J. de Phys. 34, 35-48 (1973).

9. U. Efron, et al., "The silicon liquid-crystal light valve," J. App. Phys. 57, 1356-1368 (1985).

10. I. N. Kompanets, A. V. Parfenov, and Y. M. Popov, "Spatial modulation of light in a photosensitive structure composed of a liquid crystal and an insulated gallium arsenide crystal," Sov. J. Quant. Elect. 9, 1070-1071 (1979).

11. Y. Takami and F. Shiraishi, "Surface chemical treatment effects in ultra-high purity p-type Si detectors," IEEE Trans. Nuc. Sci. NS-30, 376-386 (1983).

12. D. Armitage, W. W. Anderson, and T. J. Karr, "High-speed spatial light modulator," IEEE J. Q. E. QE-21, 1241-1248 (1985).

13. D. Armitage, "Charge isolation in a spatial light modulator," U.S. Patent 4,619,501, Oct. 28, 1986.

14. J. S. Patel, T. M. Leslie, and J. W. Goodby, "A reliable method of alignment for smectic liquid crystals," Ferroelectrics 59, 137-144 (1984).

15. D. Armitage, "Alignment of liquid crystals on obliquely evaporated silicon oxide films," J. App. Phys. 51, 2552-2555 (1980).

16. J. Seckold and G. Zosi, "Polishing silicon surfaces strain free and flat to better than wave/20," App. Opt. 25, 3006-3007 (1986).

17. H. Mimura and Y. Hatanaka, "The use of amorphous-crystalline silicon heterojunctions for the application to an imaging device," J. App. Phys. 61, 2575-2580 (1987).

18. D. Armitage, et al., "Ferroelectric liquid crystals and fast nematic spatial light modulators," in Liquid Crystals and Spatial Light Modulator Materials, William A. Penn, ed., Proc. SPIE 684, 60-68 (1986).

Fracture of solid state lasers*

John E. Marion

University of California, Lawrence Livermore National Laboratory
P. O. Box 5508, L-490, Livermore, California 94550

ABSTRACT

The understanding of fracture in solid state lasers is reviewed, with an emphasis on fracture of high average power thin slabs. Methods for characterizing strength in the context of materials selection and device implementation are discussed. Progress in slab strengthening is analyzed; aggressive optical fabrication research (to give low damage surfaces), coupled with the implementation of existing schemes for placing the slab surfaces in compression, are recommended to reduce the problem of slab fracture.

1. INTRODUCTION

With the renaissance in solid state lasers, particularly thin slabs for high average power, understanding the slab laser fracture problem has become essential.[1] Descriptions of the fracture morphology in slab lasers and the fracture criterion for brittle laser materials have been given,[2,3] and significant progress in strengthening of glass and crystalline laser media has been reported.[4-7] A review of the meaning and use of strength in the two steps of laser design, materials selection and device implementation, have also been developed.[8]

Here the state of knowledge in fracture of solid state lasers is reviewed with recommendations on future areas for fruitful research.

The fracture problem in slab lasers is a consequence of thermal stresses. The temperature of the active media in a laser amplifier rises by absorption of pumplight.[1] Generally, the surfaces of the gain element are cooled by a liquid or gas resulting in a temperature gradient: hot in the center, cool on the surfaces. This steady-state thermal gradient gives tensile stresses at the surfaces. This is an unfortunate situation for brittle materials since these surfaces are almost always the location of the strength limiting defects which result from optical fabrication.[2] When the thermal gradient becomes large enough that the resulting tensile surface stress exceeds the fracture strength, the component ruptures catastrophically.

2. CHARACTERIZATION OF FRACTURE TOUGHNESS AND COMPONENT STRENGTH

The meaning and use of the parameter strength in solid state lasers is the focus of a separate publication.[8] Two phases of laser design, materials selection and device implementation, require different interpretations of strength; for materials selection, an intrinsic measure of strength is required; for device implementation, an estimate of the actual slab strength is required.

2.1 Materials selection

For comparison of one material with another, the most appropriate parameters are the intrinsic material's properties. In brittle materials, strength is an extrinsic property that depends more strongly on the type and size of defects than on the intrinsic strength. To eliminate extrinsic variables, we use fracture toughness, a measure of a material's inherent resistance to crack propagation.

The foundation for modern fracture mechanics comes from Griffith,[9] who modeled a cracked solid under stress as a reversible thermodynamical process. He speculated that the free energy of a cracked plate must not increase when the crack extends; thus, the increased energy caused by the formation of new surfaces must be balanced by the work done extending the crack. Griffith's contribution was a relationship between the crack size and the force required to propagate the crack. At some critical crack size the

*Work performed under the auspices of the U. S. Department of Energy by the Lawrence Livermore National Laboratory under contract number W-7405-ENG-48.

increase in force required to increase crack length becomes negative and the crack propagates spontaneously. That level of force is characterized by the critical stress intensity factor, K_c, also known as the fracture toughness. In modern formalism, the fracture toughness uniquely relates the fracture strength for a given material to the depth of the machining flaws:[9]

$$\sigma = K_c \, (\pi a)^{-1/2} \tag{1}$$

The determination of the fracture toughness of laser materials has been reviewed.[8]

A majority of the technique development work for fracture mechanics measurements on brittle materials has been for polycrystalline ceramics. These techniques can be classified as "macrocrack" approaches wherein a single large crack in a fairly large specimen is studied. Promising recent developments for K_c characterization of small specimens are the "microcrack" techniques. Evans[10] and Lawn and co-workers[11-14] have examined several techniques. The existing analytical solutions for the stress field under the indenter (which is used to form the microcracks) are somewhat primitive. Also there is currently no protocol for validation of test results for anisotropic crystals and for materials that are not ideally brittle. However, we note that a relative measurement, while not ideal, is satisfactory for ranking a group of materials. Techniques appropriate for characterization of laser materials are now discussed.

2.1.1 Macrocrack techniques. Traditionally, K_c measurements have been performed on large precracked specimens.[10] In the context of laser materials testing, these macrocrack methods are primarily appropriate for glass where the availability of material is not a major restriction as it is for most crystals. Several macrocrack geometries have found particular favor for K_c characterization of glasses and ceramics. We recommend the double torsion specimen configuration[15] and the constant moment beam specimen[16] for macrocrack K_c measurements. These configurations have been rigorously standardized and can thus be regarded as reliable and reproducible.

2.1.2 Microcrack techniques. The characteristic cracks formed around microhardness indentations are used to measure the brittle material's resistance to crack propagation in the presence of stresses imposed by the indentation process. Lawn and co-workers[12-14] have reviewed the indentation fracture toughness measurement techniques.[12] Most common are the direct crack measurements which are taken by indenting a sample using a diamond Vickers pyramidal indenter. Above a minimum load (~1 Newton), a characteristic crack system forms in which radial cracks grow to an equilibrium length, as the indenter is unloaded. The load and crack length are then empirically related to the fracture toughness.[12]

The extremely small samples (1 mm³) required for the indentation method make it particularly attractive for the study of newly synthesized single crystals for which only small samples are available. However, characterization of a variety of glasses and single crystals using this technique have resulted in several observed difficulties.[8] In particular, anisotropy in the fracture energy in crystals leads to poorly behaved indentation behavior (from a geometrical standpoint); large cracks form along crystallographically preferred cleavage planes, whereas little or no cracking is observed in other directions. In samples for which the specimen orientation is known, this offers an immediate, semi-quantifiable means for characterizing cleavage behavior and fracture energy anisotropy. Plasticity and environmentally assisted growth of cracks to super-equilibrium lengths also create problems with interpretation of direct crack, fracture toughness data.[8]

2.2 Laser device implementation

For the component design step, the actual strength of the particular laser slab to be used is the pertinent measurement parameter. Fracture toughness gives the material's crack propagation resistance, but the actual component strength depends on the loading and on the type, orientation and size of the defects in the particular component.

One consequence of the effective ability of cracks in brittle materials to concentrate the stress is the small size of defects required to substantially weaken a component. The character and method of flaw formation during optical fabrication has been discussed elsewhere.[2] The primary point, pertinent here, is that surface flaws are very difficult to detect by any available technique and, in particular, are rarely visible by optical means since the crack opening is typically much smaller than the wavelength of visible light. Currently, there does not exist a reliable method for nondestructive characterization of the defect population in a component and no method is available for identifying the size of the worst flaw that will control the strength of the component.

One must resort to destructive techniques for characterizing the flaw distribution coupled with statistics to assign a safe operating stress, or one must conduct proof tests on each part.

A rationale for treating fracture data in order to assign a safe operating stress is required for laser design, and Weibull[17] has provided a statistical method. The specifics of this treatment of fracture data for slab laser design has been discussed.[2,3,8] In practice, representative samples are fractured, and the data are fit to an extreme valve distribution. This approach generates a Weibull plot, which illustrates the trend in likelihood of failure as a function of stress. Then appropriate operating parameters can be assigned for large brittle components on the basis of fracture data from small samples.

3. STRENGTHENING

Increasing the component strength pushes the thermal power limit for laser slabs, imposed by fracture, to higher levels.[1] Strengthening by reduction in the extent of subsurface damage[4] and by inducement of a compressive surface layer[5-7] have been demonstrated on glass laser slabs and crystals.[5,6]

In full-sized (32 cm) slab laser tests, modest strengthening has been demonstrated using a special polishing to reduce subsurface damage.[20] Strengthened laser slabs have been demonstrated using chemical ion exchange to form a compressive surface layer.[7] Of all strengthening methods, reducing the size of the defects in the component has the highest potential for increased strength; the ultimate limit being the theoretical strength of the component, about $E/10$ (where E is Young's modulus).[2] However, these components are not mechanically durable and slight damage induced by handling, for instance, dramatically reduces the strength. Compressive surface layers improve mechanical durability, but have limited potential for strengthening.[18] A combination of low subsurface damage for high strength with deep compressive layers for improved mechanical durability can give strong components in practice.[19]

It is important to note, however, that there is a limitation to the usefulness of very high strengths since the high thermal gradients required to fracture these components result in unacceptably high centerplane temperatures in which the lower laser level is thermally populated.[1] While reduced-temperature cooling schemes can combat this limitation, the current focus on moderate levels of strengthening will permit substantial improvements in the average power potential of the present architectures.

4. CONCLUSIONS AND RECOMMENDATIONS

In materials selection, the fracture toughness is used as an intrinsic measure of strength, whereas in laser implementation, a Weibull statistical estimate of the actual slab strength is used. Promising results in strengthening of slabs have been demonstrated, but emphasis should be placed on combining low damage polishing to give high strength, and compressive surface layers, to give mechanical durability. In that manner, strong, durable slabs can be realized in practice. We therefore recommend aggressive development of advanced optical fabrication methods for reduced damage, coupled with implementation of existing methods for applying surface compression.

5. REFERENCES

1. J. L. Emmett, W. F. Krupke, and J. B. Trenholme, "The Future Development of High-Power Solid State Laser Systems", Sov. J. Quantum Electron. 13, 1-23 (1983).
2. J. E. Marion, "Fracture of Solid State Laser Slabs", J. Appl. Phys. 60, 69-77 (1986); originally published as Lawrence Livermore National Laboratory, Livermore, Calif., UCRL-93543, October 1985.
3. J. E. Marion, "Fracture Mechanisms of Solid State Laser Slabs", accepted for publication in "Advances in Ceramics".
4. J. E. Marion, "Strengthened Solid State Laser Materials", Appl. Phys. Lett. 47, 694-696 (1985).
5. J. E. Marion, "Development of High Strength Solid State Laser Materials", Advances in Laser Science, Proc. Amer. Inst. Physic. 146, 234-236 (1986).
6. J. E. Marion, D. M. Gualtieri, and R. C. Morris, "Compressive Epitactic Layers on Single-Crystal Components for Improved Mechanical Durability and Strength", J. Appl. Phys., Aug. 1987.
7. K. A. Cerqua, S. D. Jacobs, B. L. McIntyre and W. Zhong, "Ion Exchange Strengthening of Nd:Doped Phosphate Glass", to be published in Proc. Boulder Damage Conf. NBS Special publication (1986).
8. John E. Marion, "Appropriate Use of the Strength Parameter in Solid-State Laser Design", J. Appl. Phys., Aug. 1987.

9. A. A. Griffith, "VI. The Phenomena of Rupture and Flow in Solids", Phil. Trans. Roy. Soc. (Lon.) 221A, 163-198 (1920).

10. A. G. Evans, "Fracture Mechanics Determinations", in Fracture Mechanics of Ceramics, Vol. 1: Concepts, Flaws and Fractography, eds. R. C. Bradt, D. P. H. Hasselman and F. F. Lange, Plenum Press, N.Y., 17-48, (1974).

11. A. G. Evans and E. A. Charles, "Fracture Toughness Determinations by Indentation", J. Amer. Ceram. Soc. 59, 371-372 (1976).

12. G. R. Anstis, P. Chantikul, B. R. Lawn and D. B. Marshall, "A Critical Evaluation of Indentation Techniques for Measuring Fracture Toughness: I", J. Amer. Ceram. Soc. 64, 533-538 (1981).

13. P. Chantikul, G. R. Anstis, B. R. Lawn and D. B. Marshall, "A Critical Evaluation of Indentation Techniques for Measuring Fracture Toughness: II, Strength Method", J. Amer. Ceram. Soc. 64, 539-543 (1981).

14. R. F. Cook and B. R. Lawn, "A Modified Indentation Toughness Technique", J. Amer. Ceram. Soc. 66, c-200-c-201 (1983).

15. A. G. Evans, J. Mat. Sci. 7, 1137 (1972).

16. S. W. Friedman, D. R. Malville and P. W. Mark, "NRL Progress", 36 (1972).

17. W. Weibull, "A Statistical Theory of the Strength of Materials", Royal Swedish Academy of Eng. Sci. Proc. 151, 1-45 (1939).

18. John E. Marion, "Fracture Mechanisms and Strengthening of Slab Lasers", New Slab and Solid State Laser Technologies and Applications, eds., S. Guch and J. Eggleston, Proc. SPIE 736, 2-12 (1987).

19. John E. Marion, "Compressive Coatings on Optical Components for Improving Mechanical Durability and Increasing Strength", Topical Meeting on Optics in Adverse Environments, Technical Digest, Optical Society of America 8, 56-59 (1987).

20. J. E. Marion, "Optimization of Laser Slab Treatments for High Strength and Good Mechanical Durability", Conference on Lasers and Electro-optics Technical Digest Series 1987 14, 116 (1987)

Optimal operation temperature of liquid crystal modulators

S. T. Wu, A. M. Lackner, and U. Efron

Hughes Research Laboratories
3011 Malibu Canyon Road
Malibu, California 90265

ABSTRACT

This paper describes analyses and confirming experiments on the optimum temperature for fast response in nematic liquid crystal (LC) modulators. It is demonstrated that the LCs or LC mixtures with higher nematic-isotropic phase transition temperatures have higher optimum temperatures and greater potential for improving the figure of merit. Also discussed is the performance of the LC mixture exhibiting optimum temperature at around room temperature.

INTRODUCTION

Nematic liquid crystals (LCs) have been utilized extensively for modulating visible radiation[1], and recently potential applications have been extended to the infrared (IR) region.[2] Two parameters which play important roles in the electro-optic application of LCs employing the phase retardation effect are response times and dynamic range. The free relaxation time (τ) of a parallel-aligned LC layer, in a small angle approximation, is determined by the visco-elastic coefficient, γ_1/K_{11} (where γ_1 represents the rotational viscosity and K_{11} the splay elastic constant) and by the LC layer thickness (d) as[3]:

$$\tau = \frac{\gamma_1 d^2}{K_{11} \pi^2} \tag{1}$$

On the other hand, the dynamic range (or available phase retardation δ) is also related to d as:

$$\delta = 2\pi d \Delta n / \lambda \tag{2}$$

where Δn is the birefringence of the LC layer and λ is the wavelength. Thus, for a given LC cell, large dynamic range and fast response times are conflicting issues. In particular, in the IR region a thick LC layer is needed in order to compensate for the increase of λ and decrease of Δn. This thick layer would lead to a relatively slow response time.

In order to select the potential LC candidates for electro-optic applications, we need to define a merit figure for LCs. In the electro-optic modulation of light using the phase retardation effect, a π phase change is usually required for achieving a high modulation efficiency and good contrast. Under this circumstance, $\tau = \gamma_1 \lambda^2 / (4\pi^2 K \Delta n^2)$; $K = K_{11}$ or K_{33} depending upon the LC alignment. For the parallel (homogeneous) aligned LC with positive dielectric anisotropy ($\epsilon_a > 0$), K represents the splay elastic constant K_{11}. On the other hand, $K = K_{33}$ (bend elastic constant) for the perpendicular (homeotropic) aligned LC with its $\epsilon_a < 0$. From the above expression for the free relaxation time of the LC director, we define a figure of merit (F.M.) as:

$$\text{F.M.} = K \Delta n^2 / \gamma_1 \tag{3}$$

for comparing or optimizing LC performance.

Several techniques have been developed for optimizing the dynamic range-response time trade-off of a LC modulator. Generally speaking, they can be grouped into four categories:

1. Improvement of intrinsic LC material properties. For example, LC or LC mixtures with high birefringence and low visco-elastic coefficient would exhibit a large merit figure. However, in practice, to achieve both high birefringence and low viscosity at the same time is a difficult task. Usually high birefringence in a LC would be accompanied by high viscosity, or vice versa.

2. Thin LC layer approach[4]. In this approach, a thin LC layer is used for speeding up the response time while sacrificing some dynamic range.

3. Dual field effect[5-9]. In this approach, one field is used to deform the LC director and another field to assist the LC relaxation. Dual frequency effect[5,6], inter-digital electric field effect[7], magnetic field effect[8,9] and optical field effect[9] are examples of the dual field effect. Although dual

field effects look promising, difficulties and drawbacks exist among each technique.

4. Temperature effect. It is known that the magnitude of both birefringence and the visco-elastic coefficient decrease as temperature rises[10]. The reduction of visco-elastic coefficient is favorable from the application standpoint; however, the decline of birefringence lowers the available dynamic range.

The purposes of this paper are (1) to analyze the optimal operating temperature of a given LC, and (2) to demonstrate that LC materials with higher nematic-isotropic phase transition temperature would have greater potential for improving the figure of merit through the temperature effect. In Section II, the optimal operating temperature of a LC cell is derived. These results are then compared with the experimental results in Section III. Finally, guidelines for choosing LCs or LC mixtures exhibiting a large temperature effect are briefly discussed.

THEORETICAL ANALYSIS

Since both Δn, γ_1 and K are all temperature sensitive, the figure of merit is expected to depend on temperature. The specific temperature behavior of these parameters is described as follows[10-15]:

$$\Delta n \sim \rho S \tag{4a}$$

$$\gamma_1 = a_1 S \, \exp(E_1/kT) + a_2 S^2 \, \exp(E_2/kT) \tag{4b}$$

$$K \sim S^2 \tag{4c}$$

where ρ represents the molecular density, $a_{1,2}$ the coefficient, S the order parameter, $E_{1,2}$ the activation energy, k the Boltzmann constant, and T the Kelvin temperature.

For most LCs, ρ varies less than 5% in the entire nematic range and can be neglected in the derivation. It should be pointed out here that Eq. (4b) describes the general temperature dependent rotational viscosity of a LC mixture[13]. It has better agreement with experimental results in the whole nematic range than the simple exponential dependence, $\gamma_1 \sim \exp(E/kT)$ [16]. For nematic LC consisting of molecules with a long conjugation chain (such as biphenyl), $a_2 = 0$ and the single activation energy representation results. On the other hand, for those LCs containing saturated bonds (such as cyclohexane and bicyclooctane), $a_1 = 0$. The LC mixtures (E-7 and E-44) we studied consist of cyanobiphenyl structure as their major components. Thus $a_2 \simeq 0$ and γ_1 is simplified to:

$$\gamma_1 \sim S \, \exp(E/kT) \tag{4d}$$

Furthermore, the temperature dependent S can be approximated well by[17,18]

$$S = (1 - T/T_{NI})^\beta \tag{5}$$

where T_{NI} stands for the nematic-isotropic phase transition temperature of LC and β is an exponent which is dependent on the LC compositions. Substituting Eqs. (5) and (4a, c, and d) into Eq. (3), the F.M. has the explicit temperature dependent form of:

$$F.M. \sim (1 - T/T_{NI})^{3\beta} \exp(-E/kT) \tag{6}$$

Eq. (6) is expected to have a maximum at $T = T_{op}$, the optimal operation temperature. This can be understood qualitatively as follows: when temperature rises the decrease in Δn is slower than the decrease in γ_1/K, resulting in an improvement in the F.M. However, when T gets close to T_{NI}, the drastic decline in Δn reduces the F.M. significantly. Thus, an optimal temperature exists for a given LC cell. To evaluate this T_{op}, we set $d(F.M.)/dT = 0$:

$$T_{op} = \frac{E}{6\beta k} \left[(1 + 12\beta kT_{NI}/E)^{1/2} - 1 \right] \tag{7a}$$

The quantity $12\beta kT_{NI}/E$ in Eq. 7(a) is small, so we can expand the square root term into power series, and

$$T_{op} \simeq T_{NI}[1 - 3\beta kT_{NI}/E + \ldots] \tag{7b}$$

From Eq. (7b), the optimal operating temperature of a LC modulator is lower than the nematic-isotropic phase transition temperature by the amount $3\beta kT_{NI}/E$. A LC with high T_{NI}, large E, and small β would have high T_{op}. The F.M. of a LC operating at T_{op} is improved by a factor of G:

$$G = \left(\frac{T_{NI} - T_{op}}{T_{NI} - T} \right)^{3\beta} \exp\left(\frac{E}{k} \frac{T_{op} - T}{T_{op} \cdot T} \right) \tag{8}$$

over that operated at temperature T. On the other hand, G^{-1} represents the normalized F.M. at temperature T. Both T_{op} and G^{-1} are verified experimentally for two commercially available LC mixtures, E-7 and E-44[19]. These two mixtures (the major compound is cyanobiphenyl, $C_nH_{2n+1}- o - o -CN$) exhibit very high birefringence, relatively low viscosity, large dielectric anisotropy, and fairly good chemical stability; thus, they are very good candidates for IR LC modulators.

EXPERIMENTAL

In the experiment, we need to measure the temperature dependent (A) birefringence and (B) visco-elastic coefficient of the LC under study.

A. Temperature dependent birefringence

The improved electrically controlled transmission technique[20] was used to characterize the temperature dependent birefringence of the E-7 and E-44 LC mixtures. A linearly polarized HeNe laser (λ = 0.633 μm) was used for the measurements. To minimize the surface effect[4], a 10 – 15-μm-thick, parallel-aligned LC cell (~1-cm-thick BK-7 glass substrates with transparent conductive coatings) was made. Results of the temperature dependent birefringence for the E-7 (dot data points; these data have been published in Ref. 12 and are included here for convenience) and E-44 (triangle data points) LC mixtures are shown in Figure 1. From the figure, Δn for both E-7 and E-44 decreases gradually with increased temperature. As is well known, a more drastic change occurs only in the vicinity of the phase transition temperatures. We may take advantage of this to improve the figure of merit, because in a certain temperature range the decrease in birefringence may be slower than that of the visco-elastic coefficient.

B. Temperature dependent visco-elastic coefficient

From the recently developed optical response time measurement technique[21], the visco-elastic coefficient of a LC can be determined easily. Results of E-7 (dots) and E-44 (triangles) are shown in Figure 2. It is found from Figure 2 that the visco-elastic coefficient of both E-7 and E-44 decreases linearly as temperature is raised. Nevertheless, this declination does not continue all the way to T_{NI}; it saturates at T_s, which is a few degrees below T_{NI}. For E-7, the turn-over phenomenon (γ_1/K bouncing back)

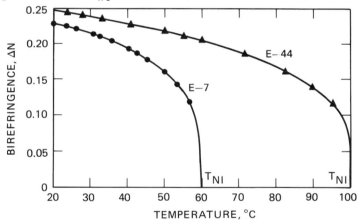

Figure 1. Temperature dependent birefringence of E-7 (dots) and E-44 (triangles) LC mixtures. The wavelength used in these measurements is λ = 0.633 μm. The T_{NI} for E-7 and E-44 is 60°C and 100°C, respectively.

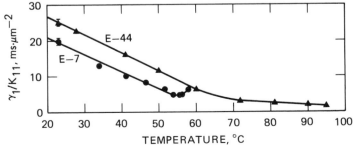

Figure 2. Temperature dependent visco-elastic coefficient of E-7 (dots) and E-44 (triangles) LC mixtures.

was observed; it implies that the decrease in K is faster than that in γ_1 as temperature approaches T_{NI}.

COMPARISON

To compare the theoretical results on the figure of merit with experiment, we need to know the values of T_{NI}, β, and E. The T_{NI} for E-7 and E-44 is 333° and 373°K, respectively. From the IR absorption dichroism measurement[18], the temperature dependent order parameter of E-7 and E-44 was obtained, and β ($\simeq 0.17$) was found to be approximately identical for both E-7 and E-44 due to their similarities in composition. Finally, the activation energy of the E-7 LC mixture was reported to be E $\simeq 0.45$ eV[22]. Although the detailed activation energy of E-44 is not available, based upon an assumed compositional similarity between E-44 and E-7, we assume that the activation energy of E-44 should be approximately the same as E-7's. Substituting these values into Eq. (7a), we obtain $T_{op} = 49.8°$ and 87.3° for E-7 and E-44, respectively. From these optimal temperatures, we can calculate the maximum F.M. for each LC. The temperature dependent F.M. can then be normalized to this maximum F.M. Results of this normalized F.M. are shown in Figures 3 and 4 for E-7 and E-44, respectively. The dots in Figures 3 and 4 represent the experimental data (normalized to the maximum value) and the solid lines represent the theoretical calculations. The agreement between experiment and theory is reasonably good. From these figures we see that the F.M. of E-7 and E-44 climb up gradually as temperature rises. In the vicinity of T_{op}, F.M. does not vary much. However, at temperatures above this region, F.M. drops drastically as the order parameter approaches zero at T_{NI}.

The absolute F.M. $(K_{11}\Delta n^2/\gamma_1)$ for E-7 and E-44 at $\lambda = 633$ nm and T = 23°C is calculated to be 2.66 and 2.40 μm^2/sec, respectively. However, the improvement factor G (as described in Eq. (8)) is more significant for the LC having higher T_{NI}. From Figures (3) and (4), G (from 23°C – 49.8°C) $\simeq 1.8$ and G (from 23°C to 87.3°C) $\simeq 6.0$ for E-7(T_{NI} = 60°C) and E-44(T_{NI} = 100°C), respectively. Thus, the maximum absolute F.M. for E-44 is about 3 times higher than E-7's at the price of operating LC at a higher temperature. The high temperature operation of a LC cell may still be tolerable even though it is somewhat inconvenient. However, for the two-dimensional liquid crystal spatial light modulator[23] (photoactivated LC light valve), the high operation temperature may introduce a certain amount of undesirable dark current in the light valve and affect its performance.

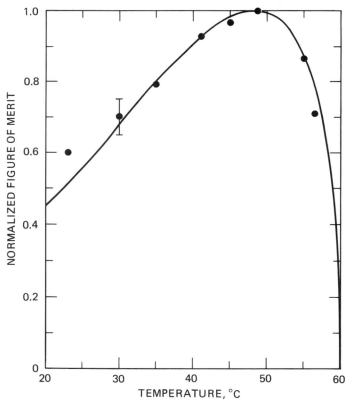

Figure 3.   Temperature dependent normalized figure of merit for the E-7 LC mixture. Dots represent the normalized experimental results and the solid line represents the theoretical calculations using Eq. (8). The parameters used for calculation are E = 0.45 eV, β = 0.17 and T_{NI} = 333°K.

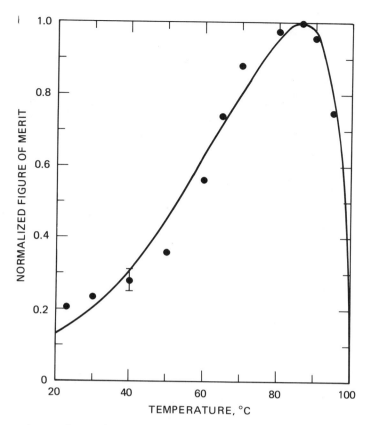

Figure 4.   Temperature dependent normalized figure of merit for the E-44 LC mixture.   Dots represent the normalized experimental results and the solid line represents the theoretical calculations using Eq. (8).   The parameters used for calculation are E = 0.45 eV, β = 0.17, and T_{NI} = 373°K.

<u>DISCUSSION</u>

As described above, the large absolute F.M. is practically important for device applications. The absolute F.M. of a LC at an arbitrary T within the nematic range can be calculated from the temperature dependent normalized F.M. if $K\Delta n^2/\gamma_1$ is known at a given temperature T_o and wavelength λ_o. Therefore, roughly speaking, LC or LC mixtures with high Δn, small γ_1/K and high T_{NI} would demonstrate large absolute figure of merit at an optimal operating temperature T_{op} as described by Eq. (7a). For a more detailed optimization, β and E need to be considered as well.

It is also interesting to examine which LC would exhibit the maximum (normalized) figure of merit at room temperature. Let us stay in the alkyl-cyanobiphenyl families, so β = 0.17 and E $\simeq$ 0.45 eV are valid. Setting T_{op} = 297°K in Eq. (7a), we find $T_{NI} \simeq$ 306°K. For this class of LC materials, their optimal operation temperature is around room temperature. Further increase in temperature can not improve the LC performance. To verify this prediction, we formulated a LC mixture, designated as HRL-7P6[24], using short chain cyanobiphenyl compounds. The T_{NI} of HRL-7P6 is ~305°K. The temperature dependent birefringence, threshold voltage, visco-elastic coefficient and figure of merit of the HRL-7P6 LC mixture have been reported in Ref. 24. The F.M. of this LC mixture does not vary much around room temperature and the maximum merit factor appears at T $\simeq$ 27.5°C, which is about three degrees higher than our calculated value. This difference may be attributed to a slightly different β or E of the HRL-7P6 mixture. Despite this small discrepancy, the absolute figure of merit of the HRL-7P6 LC is higher than many commercial LC mixtures that we have investigated at room temperature. Results of these comparisons are listed in Table I. Also included in Table I are the G and λ^* values for calculating the birefringence dispersion of LCs[12]:

$$\Delta n(T,\lambda) = G(T)\frac{\lambda^2 \lambda^{*2}}{\lambda^2 - \lambda^{*2}} \tag{9}$$

at $\lambda \gg \lambda^*$, i.e. in the IR region, $\Delta n \rightarrow G\lambda^{*2}$, which is essentially independent of wavelength except in the vicinity of molecular vibrational bands. From Table I, we find that the HRL-7P6 has the highest merit factor at room temperature. Next to HRL-7P6 is E-18, E-7,

E-44, and so on. When the room temperature operation constraint is removed, we should evaluate the absolute merit factor of LCs at their optimal operation temperatures. In this case, the T_{op} and G^{-1} of each LC have to be calculated. Although the β and E values of some LCs listed in Table I are not measured, we expect that E-44 would have the highest absolute merit factor among these LC mixtures. Results on the maximum F.M. (operating at $T = T_{op}$) for HRL-7P6, E-7 and E-44 are included in the last column of Table I.

Table I. Comparison of the merit factor for the HRL-7P6 LC and some commercially available LCs at room temperature.

LCs	Nematic Range (°C)	Δn (0.633μm)	γ_1/K_{11} (ms/μm^2)	$\Delta n^2 K_{11}/\gamma_1$ (μm^2/s)	G ($\times 10^{-6}$nm^{-2})	λ^* (nm)	$G\lambda^{*2}$	F.M. AT $T = T_{op}$
HRL-7P6	3→32	0.170	9.0	3.21	2.43	244	0.145	3.75
BDH-E18[a]	-10→60	0.218	16.5	2.88	2.82	255	0.183	---
BDH-E7[a]	-10→60	0.225	19.0	2.66	3.06	250	0.191	4.79
BDH-E44[a]	-6→100	0.245	25.0	2.40	3.53	248	0.217	14.40
RO-TN-619[b]	-10→61	0.120	6.7	2.15	2.65	204	0.110	---
ZLI-1132[c]	-6→70	0.136	10.0	1.85	3.15	198	0.123	---
RO-TN-403[b]	-10→82	0.242	34.8	1.68	3.26	250	0.204	---

a. BDH Chemicals, England
b. Hoffman La-Roche Chemicals, Switzerland
c. E. Merck, Germany

CONCLUSION

The optimal operation temperature of a LC modulator is derived. The absolute figure of merit of the E-7 and E-44 is found to be improved over room temperature operation by a factor of ~2 and ~6 by operating these LCs at their optimal temperatures, which are near their T_{NIS}. The LC mixture with its optimal temperature at room temperature was formulated. This LC mixture, indeed, exhibits a high merit figure at room temperature.

ACKNOWLEDGMENT

The authors would like to thank Drs. J. Grinberg and M. S. Welkowsky for useful discussions, Dr. J. D. Margerum for his critical review of the manuscript, and Mr. W. H. Smith and Mr. P. G. Reif for preparing the liquid crystal substrates.

REFERENCES

1. For example, see L. M. Blinov, Electro-optical and magneto-optical effects of liquid crystals" (John Wiley and Sons, New York, 1983)

2. For an overview, see S. T. Wu, "Infrared properties of liquid crystals: an overview," Opt. Eng. 26, 120 (1987).

3. E. Jackman and E. P. Raynes, "Electro-Optic Response Times in Liquid Crystals," Phys. Lett. 39A, 69 (1972).

4. S. T. Wu and U. Efron, "Optical properties of thin nematic liquid crystal cells," Appl. Phys. Lett. 48, 624 (1986).

5. H. K. Bucher, R. T. Klingbiel and J. P. VanMeter, "Frequency-Addressed Liquid Crystal Field Effect," Appl. Phys. Lett. 25, 186 (1974).

6. M. Schadt, "Low-Frequency Dielectric Relaxation in Nematics and Dual-Frequency Addressing of Field Effects," Mol. Cryst. Liq. Cryst. 89, 77 (1982).

7. D. J. Channin and D. E. Carlson, "Rapid Turn-off in Triode Optical Gate Liquid Crystal Devices," Appl. Phys. Lett. 28, 300 (1976).

8. H. J. Deuling, E. Guyon and P. Pieranski, "Deformation of Nematic Layers in Crossed Electric and Magnetic Fields," Solid State Commun. 15, 277 (1974).

9. S. T. Wu, "Dual Field Effect on Liquid Crystal Relaxation," J. Appl. Phys. 58, 1419 (1985).

10. P. G. deGennes, The Physics of Liquid Crystals (Clarendon, Oxford, 1974).

11. E. M. Aver'yanov and V. F. Shabanov, "Structural and optical anisotropy of liquid crystals," Sov. Phys. Crystallogr. 23, 177 (1978).

12. S. T. Wu, "Birefringence dispersions of liquid crystals," Phys. Rev. A 30, 1270 (1986).

13. V. V. Belyaev, S. A. Ivanov and M. F. Gorebenkin, "Temperature dependence of rotational viscosity of nematic liquid crystals," Sov. Phys. Crystallogr. 30, 674 (1985).

14. A. Saupe, Z. Naturforsch 15a, 810 (1960).

15. T. C. Lubensky, "Molecular description of nematic liquid crystals," Phys. Rev. A2, 2497 (1970).

16. H. Imura and K. Okano, "Temperature Dependence of the Viscosity Coefficients of Liquid Crystals," Jpn. J. Appl. Phys. 11, 1440 (1972).

17. I. Haller, "Thermodynamic and static properties of liquid crystals," Progr. Solid State Chem. 10, 103 (1975); (edited by McCaldin and Somorjai, Pergamon Press, Oxford).

18. S. T. Wu, "IR markers for determining the order parameters of uniaxial liquid crystals," (the accompanying paper).

19. BDH Chemical Ltd., Poole Dorset, England BH12 4NN.

20. S. T. Wu, U. Efron and L. D. Hess, "Birefringence Measurements of Liquid Crystals," Appl. Opt. 23, 3911 (1984).

21. S. T. Wu, "Phase retardation dependent optical response time of parallel-aligned liquid crystals," J. Appl. Phys. 60, 1836 (1986).

22. M. Schadt and F. Muller, "Physical Properties of New Liquid Crystal Mixtures and Electrooptical Performance in Twisted Nematic Displays," IEEE Trans. Electron Devices, ED-25, 1125 (1978).

23. U. Efron, J. Grinberg, P. O. Braatz, M. J. Little, P. G. Reif, and R. N. Schwartz, "The Silicon Liquid Crystal Light Valve," J. Appl. Phys. 57, 1356 (1985).

24. S. T. Wu, A. M. Lackner, W. H. Smith and U. Efron, "Guidelines for selecting and synthesizing nematic liquid crystals," Proc. SPIE 684, 69 (1986).

ADVANCES IN NONLINEAR POLYMERS AND INORGANIC CRYSTALS,
LIQUID CRYSTALS, AND LASER MEDIA

Volume 824

Session 2

Molecular and Polymeric Optoelectronic Materials I

Chair
Anthony F. Garito
University of Pennsylvania

Invited Paper

Total Internal Reflection Spatial Light Modulator Using Organic Electrooptic Crystal MNA

Z. Wen
Department of Electrical Engineering

C. Grossman and A.F. Garito
Department of Physics and
Laboratory for Research on the Structure of Matter

N.H. Farhat
Department of Electrical Engineering and
Laboratory for Research on the Structure of Matter

University of Pennsylvania
Philadelphia, PA   19104

ABSTRACT

A total internal reflection (TIR) spatial light modulator (SLM) employing the organic electrooptic crystal 2-methyl-4-nitroaniline (MNA) is discussed. The design configurations and performance characteristics are presented and compared with devices using inorganic electrooptic crystals such as $LiNbO_3$. It is estimated that the diffraction efficiency of a TIR SLM using MNA is enhanced by a factor of more than six compared with that using $LiNbO_3$, and, correspondingly, the driving voltage of the very large scale integration (VLIC) circuit is reduced approximately by a factor of two and half.

I. INTRODUCTION

The total internal reflection spatial light modulator (TIR SLM) is a very attractive one-dimensional electrooptic modulator that images a single scan line comprised of a large number of resolvable image elements (pixels)[1-6]. Each pixel is electronically addressed by a VLIC integrated circuit and can be updated in a random access pattern just as easily as in a linear sequence. By combining with an optical deflector, such as a rotating polygon mirror, or an acoustooptic deflector, the TIR SLM can be used for laser printing applications. It can also be used directly as one-dimensional coherent optical transducer, an electronically programmable spatial frequency filter, and in other applications.

Spatial light modulators (both one-dimensional and two-dimensional) developed up to the present time, have basically relied on inorganic electrooptic crystals such as $LiNbO_3$, $Bi_{12}SiO_{20}$ and KD_2PO_4. This is due to the availability of high quality crystals and well-established data about these materials. There has been, however, an increasing interest recently in the unusually large nonlinear optical properties of organic and polymeric solids. In particular, 2-methyl-4-nitroaniline (MNA) has received considerable attention for its exceptionally large figures of merit for second harmonic generation and linear electrooptic properties[7]. Because of its unique, polar aligned noncentrosymmetric molecular packing[8-10], the MNA crystal exhibits a large linear electrooptic effect (Pockel's effect), which is even larger than that of $LiNbO_3$[9-10]. In this paper, we propose a TIR SLM design which employs the MNA crystal. In the following section, a brief review of MNA's properties and parameters is provided. The modulator configurations and performance analysis are then described in section III. Finally, we summarize our results in section IV.

II. REVIEW OF MNA's PROPERTIES

MNA is a noncentrosymmetric substituted benzene derivative. Good quality crystals are obtained both by solution growth methods and vacuum sublimation techniques[8]. The crystals appear bright yellow and have a definite monoclinic platelet structure with b-axis perpendicular to the platelet plane. Typical solution growth crystals have dimensions of approximately 1x1x0.2 cm^3. X-ray studies determined that the MNA crystals belong to the space group Cc and point group m[9-10]. The principal axes of the index ellipsoid have been determined under a polarizing microscope[8-10]. The Y principal axis, which is fixed by the crystal symmetry, is perpendicular to the as-grown platelet. The X and Z axes are in the as-grown flat plate with X axis approximately parallel to one crystal edge. Both electrooptic and electroreflectance studies[9-11] show that only one electrooptic coefficient, r_{11}, is especially large and the following relationship among values of r_{ij} holds: $r_{11} \gg r_{31} \gg r_{13} \gg r_{33}$. The MNA crystal is nearly transparent over a range from 1900 nm to less than 500 nm along the Y principal axis, which makes a helium-neon laser at 632.8 nm well within its transmission window. Table 1 lists all the parameters of MNA which we will use in the next section.

Table 1: Parameters of MNA

$r_{11}(10^{-12}m/v)$	n_x(a)	n_y(a)	n_z(a)	ε_x(b)	ε_y(b)
67±25	2.356	1.774	1.452	5.02	3.89

(a) refractive index at λ= 632.8nm   (b) low frequency dielectric constant

III.   DEVICE CONFIGURATIONS AND PERFORMANCE ANALYSIS

The basic structure of TIR SLM has been discussed in detail in Reference 5. Figure 1 is three expanded views of the device. The device consists of a VLIC integrated circuit having serial-to-parallel addressing electronics that provide data to several thousand drive transistor which then connect to several thousand parallel metal lines on the surface of the chip. This chip is pressed against an electrooptic crystal such as MNA so that fringing fields existing between metals lines are proximity coupled into the crystal (Figure 2). These fringing fields change the index of refraction of the crystal through the Pockel's effect. To interact with these fringing fields, the device is illuminated with a sheet of light which is collimated in the horizonal axis and brought to a sheet focus on the proximity coupled surface. The light is totally internally reflected at the surface. As the light approaches this surface and recedes from this surface, it picks up a phase shift dependent on the index of refraction change produced by the fringing electric fields. By using a standard Schlieren imaging system, this phase modulated profile can be converted into an optical image intensity profile.

Two components of electrical field exist inside the modulator crystal, namely, the orthogonal field which is perpendicular to the coupling surface, and the tangential field which is parallel to the coupling surface but orthogonal to the metalic electrodes (Fig. 2). Which of the two field components drives the index of refraction change depends on the modulator crystal and its orientation. Since r_{11} is much larger than the other electrooptic coefficients for MNA, a tangentially driven modulator occurs in MNA when the crystal's X-axis is parallel to the tangential component of the electrical field, which corresponds to the as-grown crystal cut, namely a Y-cut crystal. Similarly, an orthogonally driven modulator occurs when the crystal's X-axis is parallel to the orthogonal component of the electrical field, which corresponds to a X-cut crystal.

A coordinate system (x,y,z) is defined such that the z axis is normal to the proximity coupling surface, the y axis is parallel to the metal electrodes, and the x axis is parallel to the proximity coupling surface but orthogonal to the electrode length. In most cases, the TIR SLM is operated in the Raman-Nath regime of the light coupling. It assumes that light travels in straight line inside the modulator crystal approximately along the y-axis (i.e., along the electrodes) with a small grazing angle θ. This light coupling model is valid when the grazing angle θ is not so small that Bragg regime phenomena can significantly affect the optical process inside the modulator crystal. Then the optical phase shift of light, $\psi(x)$, after its approaching and receding from the proximity coupling surface is

$$\psi(x) = \frac{2\pi}{\lambda} \int_{\frac{1}{2}L}^{\frac{1}{2}L} \Delta n_x(x,z)dl = \frac{2\pi n_x^3 \pi r_{11}}{\lambda} \int_{\frac{1}{2}L}^{\frac{1}{2}L} E(x,z=\theta|y|)dy \qquad (1)$$

where λ is the light wavelength, n_x is the index of refraction along the X principal axis, r_{11} is the electrooptic coefficients of MNA and $E(x,z)$ is coupling electrical field inside the modulator crystal. It can be shown[5] that in the limit of infinite electrode length $L \rightarrow \infty$, the optical phase shift $\psi(x)$ for an orthogonally driven TIR SLM is simply given by

$$\psi(x) = \frac{2\pi n_x^3 r_{11}}{\lambda\theta} \varphi(x,z=0) \qquad (2)$$

and for a tangentially driven TIR SLM

$$\psi(x) = \frac{2\pi n_x^3 r_{11}}{\lambda\theta} \varphi_H(x,z=0) \left(\frac{\varepsilon_y}{\varepsilon_x}\right)^{1/2}$$

In equation 2, θ is the grazing angle of the incident light, ε_x and ε_y are the dielectric constant of MNA along the X and Y principal axes, respectively; $\varphi(x,z=0)$ is the electrostatic potential on the proximity coupling surface; and $\varphi_H(x,z=0)$ is the Hilbert transform of $\varphi(x,z=0)$ defined by

$$\varphi_H(x,z=0) = \frac{1}{\pi} \int_{-\infty}^{+\infty} \frac{\varphi(x',z=0)}{(x-x')} dx' \qquad (3)$$

The proximity coupling surface potential $\varphi(x,z=0)$, which is independent on the modulator crystal[12], can be described as

$$\varphi(x,z=0) = \sum_k V_k \; \varphi_{unit} \; (x-kp,z=0) \qquad (4)$$

where p is the distance between electrode centers, V_k is the voltage placed on the kth electrode and $\varphi_{unit}(x-kp,z=0)$ is the surface potential when a unit voltage is placed on the electrode centered at x=0.

Previously reported TIR SLMs have used LiNbO$_3$ as the modulator crystal, and the driving voltage required to induce a radian optical phase shift in such a device has been calculated[5]. In order to make a comparison between TIR SLM device using MNA as the modulator crystal and that using LiNbO$_3$, let us take the ratio R of $\psi(x)$ for a device using MNA with $\psi(x)$ using LiNbO$_3$. Thus, for an orthogonally driven TIR SLM, the ratio R is

$$R = \frac{n_x^3 r_{11}}{n_e^3 r_{33}} \qquad (5a)$$

and for a tangentially driven TIR SLM

$$R = \frac{n_x^3 r_{11} \frac{\varepsilon_y}{\varepsilon_x}}{n_e^3 r_{33} \frac{\varepsilon_1}{\varepsilon_2}} \qquad (5b)$$

In Eq. 5, all values in the numerator are for MNA, and all values in the denominator are for LiNbO$_3$. Using the data in Table 1 for MNA and the standard values for $LiNbO_3$[13], we find that for a TIR SLM driven orthogonally, R is 2.66 and for one driven tangentially, R is 1.89.

The above results indicate that to induce the same values of optical phase modulation, the required driving voltage for a TIR device using MNA as modulator crystal is reduced by approximately a factor of two and a half, compared with that using $LiNbO_3$ as modulator crystal. Equivalently, since the optical diffraction efficiency can be approximated by $\frac{1}{2}\psi^2$ for small phase shift $\psi(x)$ for sinusoidal driving voltage pattern, under the same driving voltage pattern, the diffraction efficiency will be enhanced by a factor of more than six for a TIR device using MNA as the modulator crystal as compared with that using $LiNbO_3$ as modulator crystal.

IV. SUMMARY

A total internal reflection (TIR) spatial light modulator (SLM) using an organic electrooptic crystal has been proposed. It is shown that in the Raman-Nath light coupling regime, the driving voltage required to induce the same figure of optical phase modulation is reduced by roughly a factor of two and a half for a TIR SLM using MNA as modulator optical crystal compared with that using $LiNbO_3$, or, equivalently, under the same driving voltage pattern, the diffraction efficiency is increased by a factor of more than six. Because of the convenience of using an as-grown crystal platelet, the tangentially driven TIR SLM is an immediately attractive configuration than the orthogonally driven configuration.

ACKNOWLEDGEMENTS

It is a pleasure for one of us (AFG) to acknowledge many stimulating discussions with Drs. R. Lytel and F. Lipscomb. This research was generously supported by U.S. Air Force Offce of Scientific Research (AFOSR) and U.S. Defence Advanced Projects Agency (DARPA) Contract No. F49620-85-C-0105, National Science Foundation (Materials Research Laboratory Program) NSF/MRL Contract No. DMR-85-19059 and by NSF contract no. EET85-16685.

REFERENCES

1. R.A. Sprague, W.D. Turner, C. Koliopoulos, L. Flores and D. Heald, OSA Annual Meeting, Paper ThGl (Oct. ,1981)
2. R.A. Sprague L.N. Flores, D.L. Hecht, R.V. Johnson, A. Nafarrate, W.D. Turner, Digest of the Topical Meeting on Integrated and Guided Wave Optics,Paper FA4 (Oct.,1982)
3. D.L. Hecht, R.V. Johnson and L.N. FLores, OSA Annual Meeting, Paper FU4 (Oct. 1982).
4. R.V. Johnson and D.L. Hecht, OSA Annual Meeting, Paper FU5 (Oct. 1982).
5. R.V. Johnson, D.L. Hecht, R.A. Sprague, L.N. Flores, D.L. Steinmetz, and W.D. Turner, Optical Eng. $\underline{22}$, 665 (1983).
6. D.L. Hecht, et al., CLEO '87, Baltimore, poster paper ThK33.
7. A.F. Garito and K.D. Singer, Laser Focus $\underline{80}$, 59 (1982) and references therein.
8. C. Grossman, Ph.D. Thesis, University of Pennsylvania, 1987.
9. G.F. Lipscomb, A.F. Garito and R. Narang, Appl. Phys. Lett $\underline{38}$, 663 (1981); J. Chem. Phys. $\underline{75}$, 1509 (1981).
10. G.F. Lipscomb, Ph.D. thesis, University of Pennsylvania, 1980.
11. Y. Tokura et al., Phys. Rev. B$\underline{31}$, 2588 (1985).
12. C.S. Hartmann et al., Proc. IEEE Ultrasonics Sym., 413 (1973).
13. A. Yariv and P. Yeh, Optical Waves in Crystals (John Wiley and Son,Inc., New York,1984).

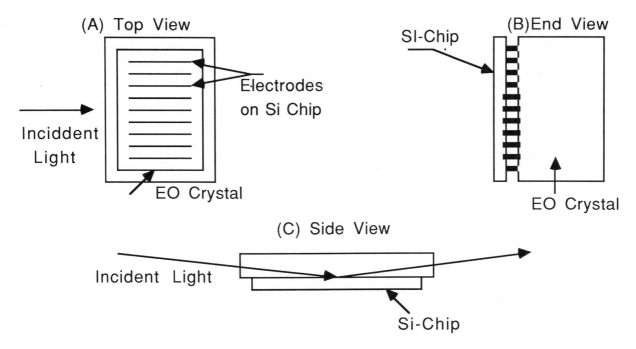

Fig.1 Three Views of The TIR Modulator: (A) Top View;

(B) End View; (C) Side View

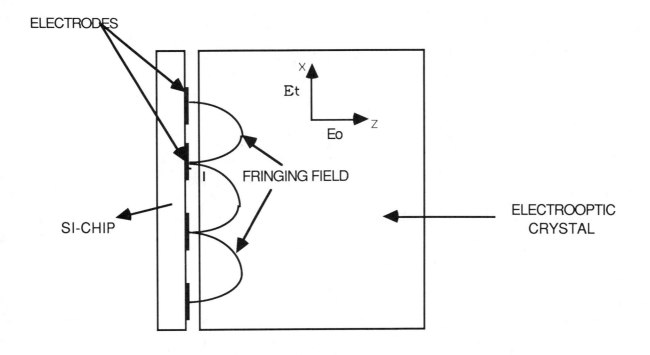

FIG.2 CRYSTAL/SI-CHIP INTERFACE AND THE FRINGING FIELD

Invited Paper

Polydiacetylene as optical memory and information processor

Eiichi Hanamura and Akira Itsubo*

Department of Applied Physics, University of Tokyo, 7-3-1 Hongo,Bunkyo-ku, Tokyo 113, and
Advanced Materials Laboratory, Mitsubishi Petrochemical Co.*, Ltd. 8-3-1 Chuo, Ani-machi,
Inashikigun, Ibaraki 300-03, Japan

ABSTRACT

Polydiacetylene crystals with different kinds of side chains are formed and show two
characteristic nonlinear-optical responses: the photo-induced structure change and the very
large third-order optical susceptibility. The former property may be used as optical
memory and the latter as the optical information processor. These two aspects of polydiacetylene are discussed in this paper.

1. INTRODUCTION

Many solids exist as metastable states at room temperature. Amorphous selenium and
tellurium systems are good examples. Irradiation of laser pulse can switch between the
amorphous and crystalline states through annealing and quenching, respectively, depending
upon the laser pulse width and power. The changes of the refractive index or transmit-
tivity of the probe light between these two states are useful as the optical memory. This
process may be called the heat mode. On the other hand, diacetylene crystal is metastable
in a sense so that it is polymerized, by ultraviolet light irradiation, into polydiacetylene crystals. For these polydiacetylene crystals, we have two isomers of acetylenic(A)
and butatrienic (B) forms. We expected[1] photo-induced isomerization between these forms by
pumping the specified electronic transition, which will be used also as materials for
optical memory. This may be called photon mode. The isomerization from A- to B- form was
confirmed experimentally[2] and the reverse isomerization was also observed, however, in the
preliminary form.[3]
We present the unified theory to understand these photo-induced structure changes.[4] It is
pointed that the photopolymerized part or photoisomerized part extends thermally or
spontaneously to the conjugation length. This property is used as amplification of optical
memory. In order to understand the microscopic mechanism of photopolymerization, we
evaluate the electronic structure of polydiacetylene and the coupling of excited electron
and hole with the lattice distortion inducing the structure change.[5]
These polydiacetylene crystals are paid attention to also because of large nonlinear
optical susceptibility. This enhancement is quantitatively understood in terms of the
calculated electronic structure. We understand that this comes from the excitons characteristic of the one-dimensional system, comparing this system with two- and three-dimen-
sional systems.

2. PHOTO-INDUCED STRUCTURE CHANGES

A model of photo-induced structure change was presented[4] to understand, from a unified
view point, mechanisms of photopolymerization of diacetylene and diolefin crystals and
photoisomerization of polydiacetylene crystal, photochromism and photo-chemical hole
burning. We can realize at the same time how a cluster of photo-excited molecules is
stabilized and how the molecular excitation far from the excited cluster is attracted to
the nearest neighbor of the cluster. This theory will give us a guiding principle to look
for and design new materials for optical memory.
We choose here the simplest model of the linear chain which is described by the
following Hamiltonian:

$$H = \frac{1}{2} \sum_i Q_i^2 - \sum_{i \neq j} K_{ij} Q_i Q_j + \sum_{\ell \in E} (E_{FC} - \sqrt{S} Q_\ell) \ . \tag{1}$$

Here Q_i describes the relevant displacement of the i-th molecule, K_{ij} the coupling constant
between the displacements of the i-th and j-th molecules, E_{FC} the Franck-Condon excitation
energy, $\sqrt{S} Q_\ell$ the Stokes shift in the excited electronic state at the ℓ-th molecule, and
$\ell \in E$ means that the summation on ℓ is taken over the sites of the excited molecules. The
first two terms of eq.(1) work in both the electronic ground and excited states. The
photo-induced structure change can not be discussed by a conventional theory of phase
transition because it is far from the thermal equilibrium. Therefore we study stability
for the cluster of m neighboring excited molecules. For this system, we rewrite the
Hamiltonian (1) in terms of the Fourier transform $Q_\ell = N^{-1/2} \sum_k e^{ik\ell} Q_k$ and in such a way as

the lattice coordinate $Q_k = Q_k^{(m)} + \Delta^{(m)}(k)$ fluctuates around the stationary displacement
$\Delta^{(m)}(k) = \sqrt{S/N}/\omega_k^2 [\sum_{\ell=1}^{m} \exp(ik\ell)]$:

$$H^{(m)} = \frac{1}{2}\sum_k \omega_k^2 Q_k^{(m)} Q_{-k}^{(m)} + mE_{FC} - \frac{1}{2}\sum_k \omega_k^2 \Delta^{(m)}(k)\Delta^{(m)}(-k) \quad . \tag{2}$$

The last term denotes stabilization energy due to the lattice distorsion. The absolute value of this energy increases by stronger than the first power of m as shown in Fig.1. Therefore under some condition, this energy overcomes the Franck-Condon energy mE_{FC} and the cluster distorted through electronic excitation becomes more stable.

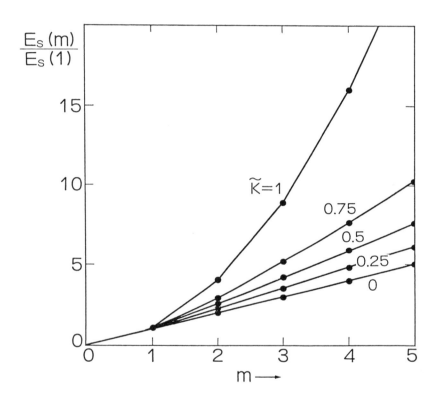

Figure 1. The distorsion energy $E_s(m)$ of a cluster of m neighboring excited molecules is plotted as a function of m. $\widetilde{K} = 4K$.

For a cluster of m neighboring molecules ($\ell = 1, 2, ..., m$) distorted through electronic excitation, the 0-th molecule is in the electronic ground state. Next we consider that the 0-th molecule is in the electronic excited state:

$$H_e^{(m+1)} = H^{(m)} + E_{FC} - \sqrt{S}Q_0 \tag{3}$$

Here we introduce the interaction mode q_0 at the 0-th molecule by

$$q_0 = \bar{\omega}\frac{1}{\sqrt{N}}\sum_k Q_k^{(m)} \quad , \quad \frac{1}{\bar{\omega}^2} = \frac{1}{N}\sum_k \frac{1}{\omega_k^2} \quad . \tag{4}$$

Then the relative stability of the electronic ground and excited states at the 0-th site is classified into the four types as shown in Fig.2. In the case of Fig.2(1), the minimum energy of the excited state is above the corresponding displaced ground state. Therefore the excited state $|e>$ at the 0-th molecule is unstable against radiative as well as nonradiative decay into the ground state $|g>$. In the case of Fig.2(2), these vertical stability at $q_0 = \sqrt{S}/\bar{\omega}$ is reversed but from the horizontal comparison, the energy-minimum

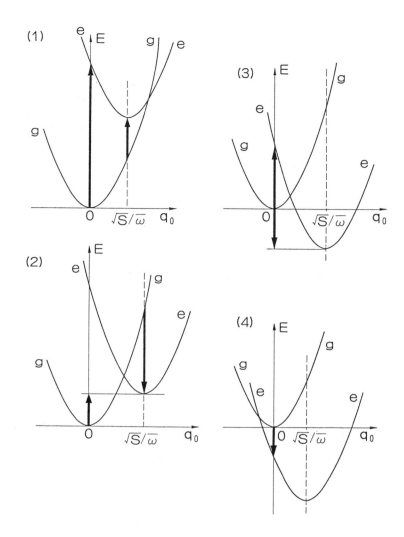

Figure 2. Energy dispersions of the 0-th molecule closest to the cluster
of excited molecules: (1) The cluster of excited molecules
is unstable, (2) metastable, (3) expands thermally and (4)
expands spontaneously without any additional optical pumping.

excited state is metastable while that of the ground state is still absolutely stable. In
the case of Fig.2(3), the horizontal stability of these states is exchanged. For a lattice
temperature higher than the potential barrier in the ground state, the molecule at the 0-th
site can jump thermally into the displaced excited state and the displaced cluster extends.
Finally in the case of Fig.2(4), the ground state at the 0-th site is unstable and changes
spontaneously into the excited displaced state. When we increase S/E_{FC}, K and/or m, the
situation proceeds from (1) to (2), (3) and (4) in Fig.2. If we fix the size m of the
displaced cluster, we have three borders as a function of $\tilde{S} \equiv S/E_{FC}$ and $\tilde{K} = 4K$; e_m between
(1) and (2), $e_m{}^*$ between (2) and (3) and g_m between (3) and (4). These borders as a
function of $\tilde{S}$ and $\tilde{K}$ are shown for several values of m in Fig.3.

One material which shows photo-induced structure change corresponds to one point on $\tilde{S}$
and $\tilde{K}$ plane of Fig.3. Then we can understand these structure changes from unified view
point. We choose polydiacetylene crystals as an example. Many kinds of polydiacetylene
crystals, e.g., TCDU, ETCD, PDA-(10-8) and -(12-8), show the photo-induced isomerization
from the acetylene (A) to the butatriene (B) types. The relevant lattice mode is such
displacements of carbon atoms as to change A-type into B-type, and the A and B phases
correspond to the ground state and the displaced excited state, respectively, in the
present model. First the yield of this A-to-B transition increases almost by the third

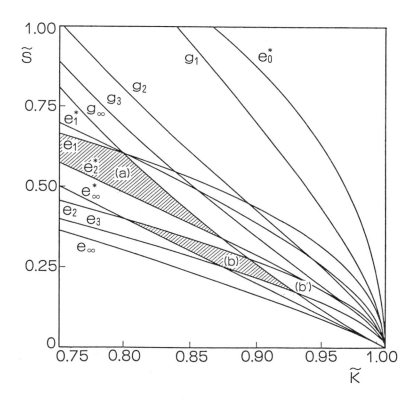

Figure 3.   The phase diagram over $\widetilde{S} \equiv S/E_{FC}$ and $\widetilde{K} = 4k$. e_m, e_m* and
g_m curves denote the stability borders of m-molecules
cluster displaced through electronic excitation. e_m: the
metastability, e_m*: the absolute stability and g_m: the
spontaneous extention of the m-displaced molecules cluster.

power[2] of the incident light intensity.   This means that only when 3 and larger than 3
neighboring molecules are optically excited, the displaced excited state beocmes stable so
that $\widetilde{S}$ and $\widetilde{K}$ are above e_3 curve and below e_2 curve. Second, the B-phase is absolutely

stable as this phase occupies the crystal after annealing in ETCD and PDA-(12-8), and
occupies the fine particles after laser irradiation at room temperature in PDA-(12-8).
Therefore the values of $\widetilde{S}$ and $\widetilde{K}$ are above e_∞ curve. Third, the photoisomerization of ETCD
and PDA-(12-8) is promptly done independently of the lattice temperature. As a result, $\widetilde{S}$
and $\widetilde{K}$ values are also above g_∞.   From these observations, we may conclude that
polydiacetylene crystals correspond to the (b') region on $\widetilde{S}$ and $\widetilde{K}$ plane in Fig.3. Note
that the amplification of optical memory is possible in the materials with $\widetilde{S}$ and $\widetilde{K}$ above g_∞
or $e*$ such as in polydiacetylenes.   In the former case, the amplification is spontaneous
while in the latter the thermal activation is required.
    Dynamics of photo-induced structure change is also studied by numerical methods. Figure
4 shows the degree of structure change N_e/N as a function of the incident light power.

This shows that we can write optical memories with the grading depending on the power of
the light pulse, when we choose a material with $\widetilde{S}$ and $\widetilde{K}$ between e_∞^* and g_∞ curves.
    For microscopic understanding of these structure changes, the band structure of poly-
diacetylene crystal is calculated by ab $initio$ tight binding approximation. The bottom of
the conduction band and the top of the valence band are exchanged for the acetylenic and
butatrienic structures against the relevant lattice distoriton Δ as shown in Fig.5.
Photo-excited electrons and holes become polarons clothed with the relevant phonons. These
polarons may be probably bound into triplet excitons with long lifetime.   Then these bound
polarons are coagulated as discussed already and induce the stable structure changes into
the B-type. This corresponds to the fact that the A-to-B structure change is induced only
when the incident photon is above the threshold of photo-conductivity.

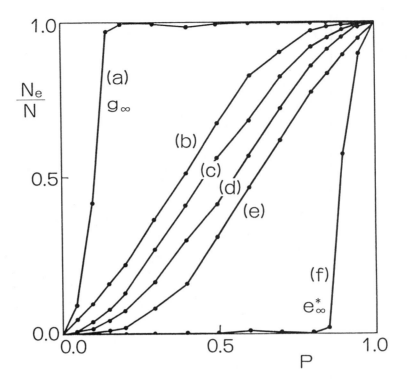

Figure 4. The final number N_e of the displaced molecules over N the total nomber as a function of normalized pump power. $\widehat{K} = 0.7$, $\widetilde{S} = 1.0346(a)$, $0.8022(b)$, $0.6550(c)$, $0.5535(d)$, $0.4792(e)$ and $0.4225(f)$.

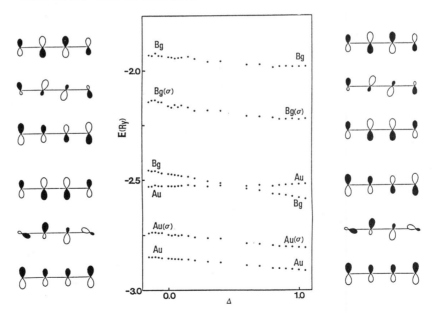

Figure 5. The change of energy levels at Γ point when lattice is deformed linearly from PDA-A ($\Delta = 0$) to PDA-B ($\Delta = 1$) configuration. The envelopes of the Bloch functions for both structure are shown, too. The envelopes of π bands are in x-y plane, and that of σ bands are in z-x plane.

3. LARGE OPTICAL NONLINEARITY

It is known that polydiacetylene crystals show large third-order optical susceptibility. To understand this origin, we obtain the electronic structure of excitons in polydiacetylene crystals and evaluate the oscillator strengths of these excitations as well as the third order optical susceptibility $\chi^{(3)}(\omega;-\omega,\omega,-\omega)$. The exciton in the bulk semiconductor has the oscillator strength reduced by $u^3/\pi a^3$, where u^3 denotes the unit cell volume and a the effective Bohr radius of the exciton. Only the states near the bottom of the conduction band and the top of the valence band within $E_{exc}^b = \mu e^4/2\hbar^2\epsilon_0^2$ are used to form the exciton, so that $u << a$ as far as the effective-mass approximation can be used, where μ is the electron-hole reduced mass and ϵ_0 the static dielectric constant. For the 2-dimensional system such as the thin quantum well of semiconductors, the electron and hole motions are confined on the plane perpendicular to the well. As a result, Coulomb attraction works more effectively so that the exciton binding energy increases to $4E_{exc}^b$ and the electron-hole separation is reduced to $a/2$. The reduction factor of the osillator strength decreases to $8u^2/\pi a^2$. The enhancement $8a/u$ comes from the increase of the exciton binding energy in comparison to 3-dimensional system and the fact that the dispersion in the well direction is missing so that whole the momentum states in this direction are used to form the exciton state. In the one-dimensional system such as polydiacetylene, the electron-hole Coulomb attraction works more effectively so that the exciton binding energy becomes infinite in the effective-mass approximation. This means that this approximation cannot be used. Therefore we use whole the states in the valence and conduction bands and the Green function method to take account of the electron-hole attraction. As a result, the one-photon allowed exciton B_{1u} has the binding energy 1.33eV and the oscillator strength 2.44 while the two-photon allowed exciton A_{1g} the binding energy 0.55eV and the oscillator strength 3.88 between A_{1g} and B_{1u} states. The conduction and valence band widths are 3.1eV and 3.5eV, respectively, so that rather large part of these states are used to form A_{1g} exciton. The electron and hole have no dispersion in two directions perpendicular to the chain so that 60% of these states are used to form the A_{1g} exciton as the sum of the oscillator strength is equal to 4 per unit cell. When we use the polydiacetylene crystal as the nonlinear optical material at the frequency 1.33eV near two-photon exciton of A_{1g} exciton, we expect $\chi^{(3)}(\omega;-\omega,\omega,-\omega)$ of an order of 10^{-8}esu when we use $\Gamma = 0.01$eV for the relaxation rate of the A_{1g} exciton.

4. CONCLUSION

Polydiacetylenes show a few characteristic nonlinear optical responses at three different frequencies: (1) the large nonlinear optical polarization at low frequency around 1.33eV, (2) the photo-induced structure change from the A- to the B-forms at the frequency above 2.3eV and (3) the reverse structure change at the ultraviolet light, although it was observed in a preliminary form. We have obtained partial theoretical understanding of these nonlinear optical phenomena.

5. ACKNOWLEDGMENTS

The authors wish to thank Akio Takada (Mitsubishi Petrochemical Co., Ltd.) for his constant encouragement throughout the present works and Yoshinori Tokura (University of Tokyo) for his fruitful discussions.

6. REFERENCES

1. E. Hanamura, Y. Tokura, A. Takada and A. Itsubo: U.S. Patents No. 835,183 (1987).
2. Y. Tokura, T. Kanetake, K. Ishikawa and T. Koda: Synthetic Metals 18, 407-412 (1987)
3. A. Itsubo and T. Kojima: private communication.
4. E. Hanamura and N. Nagaosa: Solid State Commun. 62, 5-9 (1987); J. Phys. Soc. Jpn. 56, 2080-2088 (1987).
5. H. Tanaka, M. Inoue and E. Hanamura: Solid State Commun. 63, 103-107 (1987).
6. E. Hanamura: U.S.-Japan Seminar on Quantum Mechanical Aspecs of Quantum Electronics, July 21-24, Monterey, California (1987).

Invited Paper

Characterization of polymeric nonlinear optical materials

G. Khanarian, T. Che, R. N. DeMartino, D. Haas, T. Leslie, H. T.Man, M. Sansone,
J. B. Stamatoff, C. C. Teng and H. N. Yoon

Hoechst Celanese Corporation
Robert L. Mitchell Technical Center
86 Morris Avenue
Summit, New Jersey 07901

ABSTRACT

The development of organic nonlinear optical materials requires the accurate measurement of its nonlinear optical and electrooptical properties. These measurements provide a guide to the synthesis and fabrication of new nonlinear organic materials with improved properties. A new technique for the electrooptical characterization of thin polymeric films is presented. This technique is used to measure the linear electrooptical Pockels effect in poled MNA/PMMA guest host glassy polymers. These results are compared to $\chi^{(2)}$ values derived from second harmonic generation studies. D. C. Kerr effect results are reported for a novel sol gel glass/PMMA composite material containing MNA. These results probe the relative mobility of MNA in PMMA at room temperature and elevated temperatures. $\chi^{(3)}$ values of several organic liquids are also measured by third harmonic generation studies in air. These values are compared with those derived from measurements in vacuum and good agreement is found between the two techniques.

1. INTRODUCTION

The rapid development of organic nonlinear optical materials[1,2] rests in part on the accurate evaluation of its nonlinear optical and electrooptical properties. Second harmonic generation studies have been used to infer degrees of orientation of molecules in applied electric fields (in the case of poling[3]), and to surfaces in the case of Langmuir Blodgett films.[4] In the case of guest host materials (e.g., MNA/PMMA), the temperature dependence of the D. C. Kerr effect characterizes the degree of motion of molecules in applied electric fields[16] and can be a useful complement to dielectric studies. These studies are useful for identifying optimum conditions for electric field poling of polymers. Also, the complementary study of $\chi^{(2)}$ by second harmonic generation and the linear electrooptic Pockels effect delineates the electronic from the motional contribution to $\chi^{(2)}$ which is present in polymers and guest/host materials. It has been shown that, in organic crystals[5], the electronic contribution predominates, unlike in inorganic crystals.[6] The simple and rapid characterization of $\chi^{(3)}$ of liquids and solutions by third harmonic generation allows one to evalute the third order hyperpolarizability of molecules and polymers.[7] Previous measurements[14] indicated the need to subtract the contribution of air from the observed third harmonic intensity. However, more recent work[7] indicates that accurate relative $\chi^{(3)}$ values can be obtained in air as long as the reference material is also measured in air.

This paper describes recent $\chi^{(2)}$ measurements on a MNA/PMMA poled sample to delineate the electronic contribution to the Pockels effect. The experimental results are also compared with the theoretically predicted values. The temperature dependence of the D. C. Kerr effect was also studied for the MNA/PMMA/sol gel glass composite material and compared with the results for MNA/PMMA reported earlier.[16] These studies give insight into the relative mobility of MNA in PMMA and the activation energy for the orientation process. Finally, the $\chi^{(3)}$ of several organic liquids is measured by third harmonic generation in air using the technique reported in Ref. 7, and compared with values reported in the literature, which were measured in vacuum.

2. EXPERIMENTAL METHODS AND RESULTS

2.1. Second Harmonic Generation and the Pockels Effect

The apparatus for the second and third harmonic generation studies is shown in Fig. 1. The dye laser is pumped by the frequency-doubled Yag laser to give wavelengths in the range 0.5 to 0.8 μm. This is used to pump an H_2 filled Raman cell to extend the wavelength range by the stimulated Raman effect. After the appropriate wavelength is selected, it is split into two for the sample and the reference arms. The sample is rotated in the case of a slab to obtain the coherence length. The measurements are performed relative to a standard material such as quartz, BK7 glass or CS_2. The energy in each pulse is integrated by a boxcar integrator, and each pulse from the sample arm is divided by the reference arm, to compensate for shot-to-shot laser pulse fluctuations. The translation or rotation of the samples are controlled by a computer which also averages and stores data. In the case of second harmonic generation from poled polymers, the observed intensity I^{2w} is related to the fundamental pump intensity I^w by[8]

$$\frac{I^w}{(I^{2w})^2} \simeq (d_{33}l_c)^2 p^2 t_w{}^4 T_{2w} \sin^2\left[\frac{2\pi L}{\lambda}(n_w\cos\theta_w - n_{2w}\cos\theta_{2w})\right]$$

(1)

In the above equation t_w and T_{2w} are Fresnel transmission factors[9] which account for the angular dependence of the transmission of the linear and nonlinear optical waves, d_{33} is the nonlinear optical coefficient along the poling direction, l_c is the coherence length, L is the thickness, n is the refractive index, θ is the angle of rotation and p is a projection factor of the optical electric field along the d_{33} and d_{31} tensor elements and also the polarization vector. The explicit expression is given in Ref. 8. Fig 2 shows a typical arrangement for the measurement of second harmonic from a poled film. Fig. 3 shows a typical Maker fringe pattern for a poled polymer. At normal incidence, there is no second harmonic intensity because there is no projection of optical electric field along a d tensor element. d_{33} is related to $\chi^{(2)}$ by the relation $\chi^2_{zzz}(-2w,w,w) = 2d_{33}$.

The linear Pockels effect is measured by an apparatus shown in Fig. 1 in Ref. 16. However, because one is dealing with very thin samples of the order 2 to 200 μm thick, one cannot easily measure the electrooptic properties by transmitting light along the length of the sample since the light would have to be waveguided along it. Instead we adopt the measurement geometry shown in Fig. 4. The light is transmitted along the thickness of the sample, with its polarization at 45 degrees to the horizontal or vertical planes. The electric field is applied across the thickness of the sample. It can be shown[10] that, for a poled film with axial symmetry about the poling field, only the transverse Pockels effect is possible. Hence, at normal incidence, the light detects no birefringence. The angular dependence of the birefringence is given to a good approximation by

$$\Delta n(\theta) = \Delta n(\theta = 90°) \sin^2\theta$$

(2)

where $\Delta n(\theta = 90°)$ is the birefringence one would measure if the light traversed the length of the sample, and θ is the angle of incidence in the sample. As the sample is, rotated the optical path length also increases by

$$l(\theta) = \frac{d}{\cos\theta}$$

(3)

where d is the thickness of the film. Thus the optical retardation ϕ is given by

$$\phi = \frac{2\pi}{\lambda} d\Delta n(\theta = 90°)\frac{\sin^2\theta}{\cos\theta}$$

(4)

where λ is the wavelength of light. The above approximate expression holds in the limit of small induced birefringence. The phase retardation is measured by applying sinusoidal voltages across the sample and using synchronous techniques to detect the modulated optical signal. The measurements are usually done at a fixed angle relative to the laser beam and from Eqn. 2, $\Delta n(\theta = 90°)$ is deduced. This is related to the Pockels coefficients by

$$\Delta n(\theta = 90°) = \frac{1}{2}(n_3^3 r_{33} - n_1^3 r_{31})\frac{V}{d}$$

(5)

where r_{33} and r_{31} are the Pockels coefficients corresponding to the diagonal and offdiagonal terms and n_3 and n_1 are the corresponding refractive indices. Usually the birefringence is not high and $n_3 = n_1$. Also one can show that $r_{31} = r_{33}/3$.[3] Having evaluated r_{33}, it is simply related to $\chi^{(2)}$ by

$$\chi^{(2)}_{zzz} (-w,w,o) = \frac{-1}{2} \varepsilon n^2 r_{33}.$$

Fig. 5 shows a typical plot of the output of the detector versus the applied electric field and the linear dependence verifies that the Pockels effect is being measured.

The second harmonic generation study on 10% MNA/PMMA was done with 1.06 μm radiation. The Pockels effect study was done using 0.63 μm light and a dispersion relation was used[3] to convert the $\chi^{(2)}$ to 1.06 μm. Table 1 gives the $\chi^{(2)}$ values obtained from second harmonic generation and the Pockels effect. In the case of the Pockels effect, there can be both motional and electronic contributions to $\chi^{(2)}$ whereas, in the case of $\chi^{(2)}$ derived from second harmonic generation, only the electronic contributions are important. The fact that the $\chi^{(2)}$ values from the two experiments agree closely shows that the electronic contribution predominates in the Pockels effect also at the frequency at which the experiment was performed (2000 Hz). This means that the rotational diffusion time constant in an electric field of the MNA molecule in PMMA must be considerably greater than 0.5 milliseconds. This result has important implications for the utility of organic $\chi^{(2)}$ materials in that they are virtually dispersionless from D. C. to optical frequencies. It is also useful to predict the $\chi^{(2)}$ values of polymers based on theoretical models. A model has been derived on the basis of molecules being free to rotate in the applied electric field independent of one another, subject to the Boltzmann statistics. The model predicts that[3]

$$\chi^{(2)}_{zzz} = Nf^3 \frac{\varepsilon (n^2+2)}{n^2+2\varepsilon} \frac{\mu \beta E}{5kT} \tag{6}$$

where μ is the dipole moment, β is the hyperpolarizability, E is the poling field, k is the Boltzmann constant, and T is the temperature. ε and n are the dielectric constant and refractive index, respectively. f is an internal field factor. Evidence for the correctness of the oriented gas model is given by the temperature dependence of the Kerr effect, which will be discussed below. Nevertheless, the agreement between the experimental and theoretical calculations is not good (see Table 1). The origin of the discrepency may lie in the value used for β in Eqn. 6. We used the value reported by Teng and Garito[11] for an isolated MNA molecule. However, MNA in PMMA finds itself in a dense medium of dielectric constant of 3 and a solvatochromic effect[11] must be present which increases the apparent β of MNA. Indeed, when MNA is dissolved in dioxane (ε = 2), its β value increases; thus better agreement may be obtained between the theoretical and experimental $\chi^{(2)}$ values.

2.2. D. C. Kerr Effect

The Kerr effect of the MNA/PMMA/solgel glass composite was also measured by the same apparatus as for the Pockels effect. The only difference was that the samples were thick enough (3mm) for light to pass along the length of the sample and the Kerr constant relative to CS_2 was measured. The methods of fabrication of the sample are described elsewhere[12]. The samples were heated and the Kerr constant measured as a function of temperature. Fig. 6 is a plot of the Kerr constant normalized by its value at 298°K, versus the inverse of the temperature.

The temperature dependence of the Kerr effect of the 5% MNA/PMMA/solgel glass composite is interesting for a number of reasons. Firstly, the Kerr constant increases exponentially between room temperature and 50°C. This indicates that the MNA finds itself in a restricted environent and is limited in its response to the electric field by the surrounding environment. Indeed, the activation energy for this process is about 28 kcal/mole, which corresponds approximately to the activation energy of PMMA chain mobility. There is also good agreement between the results in the sol gel glass matrix and those reported earlier[16] for MNA/PMMA alone. This implies that the sol gel glass is acting simply as a support medium but that the MNA is in molecular proximity to the PMMA only. Above 50°C, the data is interesting because the sol gel glass matrix enabled us to go to higher temperatures without the material softening and losing its optical properties, in contrast to MNA/PMMA, which softens above 50°C and loses its good optical quality. The data on the sol gel glass system indicates that the Kerr constant reaches a plateau which, in fact, corresponds to the value it would have if it were in a free solution[16]. These data also support the view that, at elevated temperatures when MNA/PMMA is poled, it is relatively free to orient in an applied electric field and should obey the Boltzmann statistics from which Eqn. 6 is derived.

2.3. Third Harmonic Generation

Third harmonic generation studies were carried out with the same apparatus as shown in Fig. 1. Liquids were placed in glass wedge cells and translated across the laser beam. Fig. 7 shows a typical Maker fringe plot from which the coherence length and amplitude are measured. These studies were carried out in air, which contributes its own third harmonic polarization. Other workers have performed experiments in vacuum[13,14] to eliminate that contribution. However, recently it has been shown[7] that accurate relative measurements can be performed in air, and we have followed that proceedure.

We have used Eqn. 18 in Ref. 7 to calculate the $\chi^{(3)}$ of a number of liquids from our third harmonic generation measurements, which are listed in Table 2. We have also listed values reported in the literature that were measured either by evacuating the air around the wedge cell or by explicitly taking into account the contribution of the air. We note that there is good agreement between the two columns. As pointed out in Ref. 7, the reason why measurements can be carried out in air is because it contributes the same phase factor both to the standard CS_2 and the sample being measured. Thus a relative measurement of $\chi^{(3)}$ materials is possible in air as long as the standard is also measured in air.

3. CONCLUSIONS

We have shown that electrooptical and nonlinear optical measurements provide useful insights into the molecular and material properties of organic nonlinear optical materials. From Pockels electrooptical and second harmonic generation measurements, it has been shown that $\chi^{(2)}$ is essentially without dispersion, because of the dominant electronic contribution to the Pockels effect. The D. C. Kerr effect measurements on MNA/PMMA/solgel glasses show that MNA is hindered in its motion at room temperature but that, at elevated temperatures, it has liquid-like freedom. Third harmonic generation studies were performed on a number of liquids in air and good agreement was found with those done in vacuum. Thus we have demonstrated a relative method for characterizing materials in air by third harmonic generation.

4. ACKNOWLEDGMENT

We wish to thank Dr. Don Ulrich of AFOSR and Dr. John Neff and DARPA for their support and encouragement in carrying out this work. This work was carried out under Contract Number F49620-86-C-0129.

Table 1. $\chi^{(2)} \times 10^{-9}$ esu for 10% MNA/PMMA at 1.06 μm.

$\chi_{zzz}^{(2)}$ $(-2w,w,w)$	$\chi_{zzz}^{(2)}$ $(-w,w,o)$	$\chi_{zzz}^{(2)}$ (calc.)
1.6	1.5	0.6

Table 2. $\chi^{(3)} \times 10^{-14}$ esu for Several Organic Liquids

Liquid	$\chi^{(3)}$	
	Present Study	Literature[14],[15]
CS$_2$	23	standard
Cyclohexanone	4.4	
Hexane	4.0	5.3
Methylene chloride	5.7	
Dimethylsulphoxide	7.8	
Dimethylformamide	5.3	5.05
Dimethylacetamide	6.8	
Toluene	11	9.8

5. REFERENCES

1. J. B. Stamatoff, A. Buckley, G. Calundann, E. W. Choe, R. N. DeMartino, G. Khanarian, T. M. Leslie, G. V. Nelson, D. Stuetz, C. C. Teng, and H. N. Yoon, Proc. SPIE, Vol. 682, p. 85 (1987).
2. R. N. DeMartino, E. W. Choe, D. Haas, G. Khanarian, T. Leslie, G. V. Nelson, J. B. Stamatoff, D. Stuetz, C. C. Teng and H. N. Yoon, ACS Denver Symposium, Plenum, in press (1987).
3. K. D. Singer, M. G. Kuzyk and J. E. Sohn, J. Optical Soc. B., Vol. 4, p. 968 (1987).
4. G. Khanarian, Thin Solid Films, 152, p. 265 (1987).
5. F. G. Lipscomb, A. F. Garito and R. S. Narang, J. Chem. Phys., 75, p. 1509 (1981).
6. M. E. Lines and A. M. Glass, "Principles and applications of ferroelectrics and related materials", Clarendon Press, Oxford (1979).
7. F. Kajzar, and J. Messier, J. Optical Soc. B, Vol. 4, p. 1040 (1987).
8. K. D. Singer, J. E. Sohn and S. J. Lalama, Appl. Phys. Lett., 49, p. 248 (1986).
9. J. Jerphagnon and S. K. Kurtz, J. Appl. Phys., 41, p. 1667, (1970).
10. D. Raskin and G. Khanarian, Celanese internal technical memorandum (1987).
11. C. C. Teng, and A. F. Garito, Phys. Rev. B., 28, p. 6766 (1983).
12. T. Che, R. Carney, G. Khanarian, R. Keosian, and M. Borzo, J. Noncrystalline Solids, in press (1988).
13. S. Stevenson, and G. R. Meredith, Proc. SPIE, Vol. 682, p. 147 (1987).
14. G. R. Meredith, B. Buchalter, and C. Hanzlik, J. Chem. Phys., 78, p. 1543 (1983).
15. F. Kajzar and J. Messier, Phys. Rev. A, 32, p. 2352 (1985).
16. G. Khanarian, A. Artigliere, E. W. Choe, R. N. DeMartino, R. Keosian, D. Stuetz and C. C. Teng, Proc. SPIE, Vol. 682, p. 153 (1987).

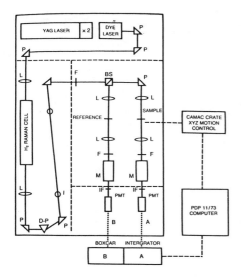

Fig. 1 Schematic of laser apparatus.

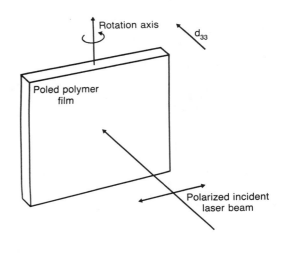

Fig. 2 Experimental geometry for measuring second harmonic generation from poled film.

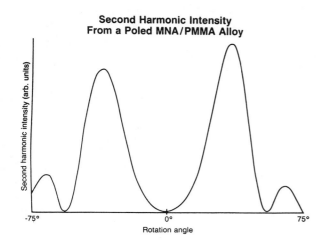

Fig. 3

Rotation
Axis
r_{33}
Poled Polymer
Film
r_{31}
θ
45°
Polarized
Incident
Laser Beam

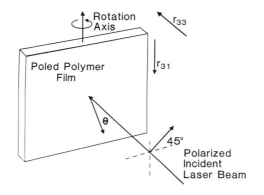

Fig. 4 Experimental geometry for measuring Pockels effect in thin film.

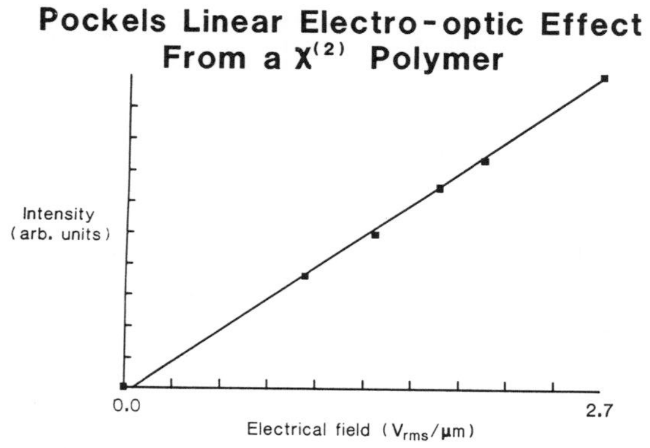

Pockels Linear Electro-optic Effect
From a $\chi^{(2)}$ Polymer
Intensity
(arb. units)
0.0
2.7
Electrical field ($V_{rms}/\mu m$)

Fig. 5

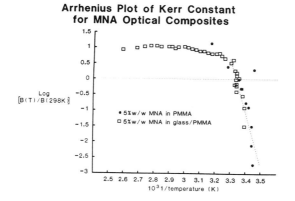

Arrhenius Plot of Kerr Constant
for MNA Optical Composites
Log
[B(T)/B(298K)]
1.5
1
0.5
0
-0.5
-1
-1.5
-2
-2.5
-3
• 5%w/w MNA in PMMA
□ 5%w/w MNA in glass/PMMA
2.5 2.6 2.7 2.8 2.9 3 3.1 3.2 3.3 3.4 3.5
10^3 1/temperature (K)

Fig. 6

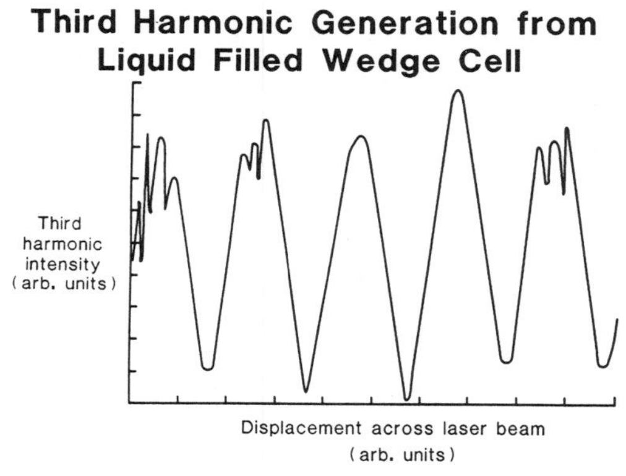

Third Harmonic Generation from
Liquid Filled Wedge Cell
Third
harmonic
intensity
(arb. units)
Displacement across laser beam
(arb. units)

Fig. 7

Nonlinear Optics and Electro-Optics in Langmuir-Blodgett Films

G H Cross, I R Peterson and I R Girling

GEC Research Limited, Hirst Research Centre, East Lane,
Wembley, Middlesex HA9 7PP

ABSTRACT

The electro-optic response of nonlinear supermolecular arrays of organic dyes, prepared by the Langmuir-Blodgett (LB) technique is measured by an attenuated total reflection (ATR) method. The field induced dielectric constant changes in the films modify the coupling conditions for the excitation of a silver surface plasmon mode and produces a differential reflectivity in the region of the ATR minimum. The magnitude of the response is measured for film thicknesses corresponding to 1 through to 13 monomolecular LB layers. An anomalous dependence is seen. The anomaly occurs at a film thickness which segregates two thickness regimes for which the differential responses are qualitatively dissimilar. Structural re-organisations in the thin film regime may be responsible.

1. INTRODUCTION

The high frequency response of certain organic materials are at least comparable to some of the inorganic nonlinear glasses (e.g. Lithium Niobate, $LiNbO_3$) and because of the ease of chemical modification, their optical properties are readily tunable to particular applications[1,2]. It is expected that such organic materials may satisfy the requirements of lower power and higher bandwidth in devices designed for integrated optics. The Langmuir Blodgett (LB) method for the assembly of optically nonlinear organic films provides solutions to the problems of alignment and thickness control which are key considerations for nonlinear optically guiding media. The observation of second order effects in LB films of less than waveguide thickness such as second harmonic generation (SHG), is now well known[3-6]. The findings that the high frequency responses of organics are superior to some inorganics are not generally true of the (lower frequency) electro-optic responses ($LiNbO_3$, for example, benefits from ionic polarisations). Nevertheless a study of this property is important since it can point to the performance to be expected at optical frequencies. Additionally, a demonstration of the electro-optic response in these materials more easily shows their potential for inclusion in active devices.

The reliable fabrication of noncentrosymmetric LB films of organic dyes has been established[3-6] and in previous papers we have investigated the resulting electro-optic (Pockels) response in monolayer films[7,8]. In this paper we report an extension to that work in which the relationship between the magnitude of the response versus film thickness is investigated in a series of multilayer structures consisting of alternating dye layers.

2. SURFACE PLASMON POLARITONS (SPPs) AS PROBES OF NONLINEARITIES

The sensitivity of the surface plasmon-polariton as a probe of surface structure has frequently been noted[9-13]. In theory such excitations may be used to probe third order nonlinearities in media in contact with the metal surface. The optical field enhancement at the metal/non-linear dielectric interface acts through the $\chi^{(3)}$ (n_2) response to modify the refractive index and thus the SPP wavevector, such that self tuning/detuning of the resonance is possible[14-16]. The effect can be measured by following the change in specular reflectivity from the coupling medium (prism or grating). Since the reflectivity dip in such an experiment can be greater than 90% it might be possible to design an optical intensity modulator or nonlinear optical bistable switch whose bandwidth is on the order of a few hundred THz. In our previous work we have used an associated method to detect changes in the dielectric constant of nonlinear LB films arising through the Pockels response upon the application of an external electric field[7,8]. In the attenuated total reflection (ATR) geometry we can detect modulations in the reflected beam intensity which are the result of surface mode coupling/de-coupling occurring at the frequency of the applied a.c. field. Coupling of light to the surface plasmon occurs at a particular angle, θ_c, where the component of the wavevector of the incoming light parallel to the surface, k_x, matches that of the surface plasmon, k_{sp}. The experimental configuration and co-ordinate system is shown in Figure 1. The coupled optical wave is evanescent in the dielectric medium above the silver film and for an isotropic medium the electric field components E_x and E_z are related by;

$$\left| \frac{E_x}{E_z} \right| = \left| \frac{k_z}{k_x} \right| = \left| \varepsilon_1 / \varepsilon_2 \quad \sin^{-2}\theta_c \, -1 \right|^{1/2} \tag{1}$$

where the subscripts 1 and 2 refer to the infinite bounding medium and the prism, respectively.

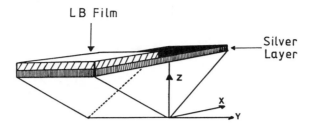

Figure 1: Sample geometry and co-ordinate system

With ε_1 taken as the linear dielectric constant of the thin LB film, equation 1 thus gives the direction of the optical electric field vector in the film. For the coupling angles in these experiments, $E_x/E_z \ll 1$ and since the principle β tensor components of these dyes are aligned towards the film plane normal (z axis), there is significant overlap between the optical field and the bulk polarisability, $\chi^{(2)}(-\omega;\omega,0)$. The applied low frequency electric field is colinear with the z axis.

The differential reflectivity, ΔR, obtained as a function of angle for the applied field, ΔE, is representative of the modifications to the complex dielectric constant, $\Delta \varepsilon$. Thus;

$$\Delta R \approx \frac{\partial R}{\partial \varepsilon^{(r)}} \, \Delta\varepsilon^{(r)} + \frac{\partial R}{\partial \varepsilon^{(i)}} \, \Delta\varepsilon^{(i)} \tag{2}$$

in which the superscripts (r) and (i) refer to the real and imaginary parts respectively. The first additive term in Equation 2 gives rise to a shift in the coupling angle, $\Delta\theta_c$, corresponding to a change in the dispersion of the SPP. The second term represents changes to the width, $\Delta\theta^{1/2}$ and depth, ΔR_{min}, of the resonance minima. The responses observed for two representative monolayers of nonlinear dye are shown in Figure 2. As a

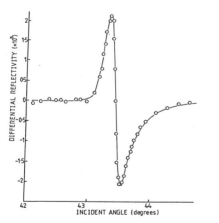

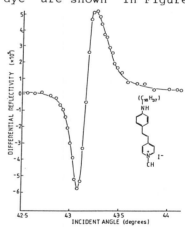

Figure 2: Differential reflectivity of ATR minima due to the electro-optic response of monolayers of (a) hemicyanine[7] and (b) N-stilbazene. The change in sense of the response is related to the relative positions of the hydrocarbon chain with respect to the direction of β in the molecules.

result of obtaining both the magnitude and phase of $\chi^{(2)}(-\omega;\omega,0)(\lim_{\Delta\varepsilon\to 0} \frac{\Delta\varepsilon}{\Delta E})$, the relative signs of the responses from different films are immediately apparent. This is a property reflecting the orientation sense of the dye. The opposition in sign between the hemicyanine film and the N-stilbazene film is a consequence of design differences between the two materials and the results of the orientational specificity of the deposition process. For monolayers, the magnitude of the response is typically, $\Delta R \approx 10^{-5}$ corresponding to a dielectric constant change, $|\Delta\varepsilon|$, of 10^{-4} at fields of $\Delta E \approx 10^7$ V/m.

Characteristic of the SPP field intensity is its exponential decay into the media on either side of the metal/dielectric interface. This is usually much longer (see Equation 1) than the thickness of a few monolayers of LB film (each monolayer is about 30Å) and thus it should be possible to benefit from the increased interaction of the field with a thicker noncentrosymmetric film. The remainder of this paper details our study of the differential reflectivity obtained from electro-optic LB films up to 13 layers (390Å) in thickness.

3. ELECTRO-OPTICS IN MULTILAYER LB FILMS

The construction of the cell used for the observation of electro-optic effects by ATR is shown in Figure 3.

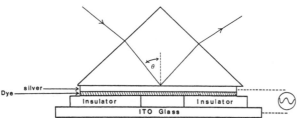

Figure 3: Cell used for ATR electro-optics.

We have utilised the Kretschmann coupled ATR experiment, modified to allow an a.c. field to be applied to the film. The evaporated silver layer in this configuration acts both as the support for the surface mode and as one of the electrodes. Fields of between 10^6 and 10^7 V/m are applied at a frequency of 4.8 kHz.

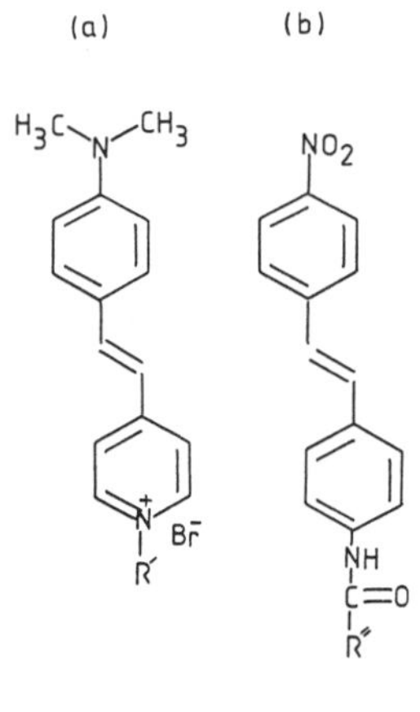

Figure 4: Chemical structure of (a) hemicyanine (R' = $C_{22}H_{45}$) and (b) 4-heptadecylamidonitrostilbene (R'' = $C_{17}H_{35}$).

The LB films consist of alternate compressed monolayers of hemicyanine (Figure 4(a)), and 4-heptadecylamidonitrostilbene (Figure 4(b)) being deposited on the upstroke and downstroke respectively in the deposition sequence. The resulting multilayer films contain 'ferroelectrically' aligned chromophore moieties giving dominant polarisability tensor components, β_{zzz} which are constructively additive[6].

Before obtaining the differential reflectivity, the characteristic parameters of the silver film were determined by observing the SPP for the prism-silver-air layer system using a HeNe laser (λ = 632.8 nm) as the probe beam, and fitting this spectrum to our reflectivity model. Treating the layers as homogeneous isotropic media, the complex dielectric constant and thickness of the silver film are obtained. Holding these parameters constant, the same experiment for the prism/silver/LB-film/air system gives a corresponding value for the complex dielectric constant of the film using a film thickness obtained from the molecular length, average tilt in each layer and number of deposited layers. Again, the isotropic model is satisfactory despite the obviously anisotropic nature of these LB films. The data fit is no better with a uniaxial model. Despite these approximations, our computed values of the real part dielectric function, $\varepsilon^{(r)}$ (Table 1) are in reasonable agreement with one another. It is of interest to note that the spread amongst the values of the imaginary part to the dielectric constant, $\varepsilon^{(i)}$, seems to be, at least in part, associated with the corresponding effective loss measured for the underlying silver layer and thus may not necessarily reflect significant variations in the intrinsic scattering between the LB films.

Table 1: The dielectric constants and differential dielectric constants obtained by the ATR spectroscopy described in the text. The figures in parentheses refer to the dielectric parameters found for the silver layer in each case.

No. of layers	$\varepsilon(r)$	$\varepsilon(i)$	$\Delta\varepsilon(r)$ x10^5	$\Delta\varepsilon(i)$ x10^5
1	2.49 (-17.17)	0.08 (0.38)	21.8	2.2
3	3.81 (-17.68)	0.16 (0.61)	27.0	-49.4
5	3.67 (-17.72)	0.43 (0.69)	56.1	-145.0
7	3.20 (-17.87)	0.03 (0.36)	18.5	-33.7
9	3.04 (-17.45)	0.00 (0.18)	4.4	-10.8
11	3.08 (-17.47)	0.15 (0.35)	24.0	165.1
13	3.12 (-17.57)	0.06 (0.39)	13.8	7.1

To obtain the differential reflectivity as a function of angle, we measure the intensity modulations in the reflected beam using a lock-in amplifier. The observed response for the 7 layer film is shown in Figure 5. Notice that the sign of the response reflects the

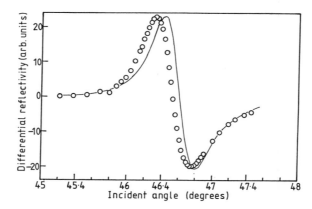

Figure 5: Differential reflectivity of the ATR minimum due to the electro-optic response of a 7 layer alternating dye multilayer. 40V peak-to-peak applied (circles). Solid line; The function $\frac{\partial R}{\partial \varepsilon(r)_\perp}$ calculated from the reflectivity.

direction of the bulk second order polarisability which is specified, in thick films, relative to the sign of $\chi^{(2)}(-\omega;\omega,0)$ in the initial monolayer. For these films, the initial layer is hemicyanine for which the corresponding response is given in Figure 2(a). That the response seen is the linear Pockels electro-optic effect is shown in the linearity of the relationship between the magnitude of the differential reflectivity and the applied voltage, for all films up to 13 layers.

When the magnitude of the response is shown against film thickness (Figure 6) we observe an apparent anomaly occuring at around 9 layers. The response falls very sharply at this point after a rise up to 7 layers. A simple idea of the expected dependence of the response versus film thickness is that the modulation will rise to a single maximum value to the point at which the confinement of further SPP field strength is traded off with a smaller differential response due to resonance broadening. We see no sudden increase in $\theta^\circledR$ at the thickness of the anomaly and both this and the coupling angle, θ_c are monotonically

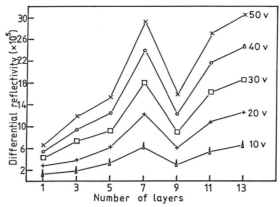

Figure 6: The peak differential reflectivity shown versus film thickness for various applied voltages.

increasing functions of film thickness.

Absolute values of the corresponding field induced changes in film dielectric constant were obtained using the functions $\frac{\partial R}{\partial \varepsilon_{\perp}^{(r)}}$ and $\frac{\partial R}{\partial \varepsilon_{\perp}^{(i)}}$ calculated numerically from the model reflectivity, R. The subscript $\perp$ signifies the dielectric tensor components normal to the film plane and thus colinear with the applied electric field. These components of dielectric constant, approximately along the chromophore axis are expected to be the most sensitive to the applied electric field. The experimental differential reflectivities may be scaled to these functions to yield the complex dielectric constant changes i.e. $\Delta\varepsilon_{\perp}^{(r)}$ and $\Delta\varepsilon_{\perp}^{(i)}$.

These parameters are shown in Table 1 where they refer to measurements made at an applied cell potential of 40V (peak to peak). The most notable feature of this table is that for the thinner films, unexpectedly, $\Delta\varepsilon_{\perp}^{(i)}$ and $\Delta\varepsilon_{\perp}^{(r)}$ are of _opposite_ sign. This applies for all films up to 7 layers whereas for the 11 and 13 layer films the response is reasonably well modelled with the assumption that the changes to the complex dielectric constant components are of the _same_ sign. The qualitative features of the 9 layer response were sample dependent. In addition the opposition in sign for the differential dielectric functions is only clearly evident at higher field strength (ie 40V applied and above). This film thus

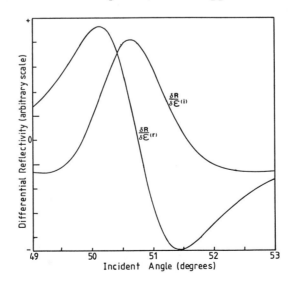

Figure 7: The structure of the functions $\frac{\partial R}{\partial \varepsilon_{\perp}^{(r)}}$ and $\frac{\partial R}{\partial \varepsilon_{\perp}^{(i)}}$ calculated for the 11 layer film.

discriminates between two thickness regimes where the qualitative features of the response differ noticeably. To best illustrate these observations we refer to Figure 7 in which the structures of the functions $\frac{\partial R}{\partial \varepsilon_\perp^{(r)}}$ and $\frac{\partial \varepsilon_\perp^{(i)}}{}$, calculated from the ATR minimum for the 11 film, are shown. The zero of $\frac{\partial R}{\partial \varepsilon_\perp^{(r)}}$ occurs at the SPP reflectivity minimum so that adding a positive multiple of $\frac{\partial R}{\partial \varepsilon_\perp^{(i)}}$ shifts the zero of the resultant trace to the <u>right</u> of the SPP minimum, and conversely to the <u>left</u> for a negative multiple. To determine the relative signs of the differential dielectric constant components, we simply note the relative direction of the shift in the experimental response (being a composite of real and imaginary dielectric modulations) away from the real part differential function. Figure 5 shows, as an example, this mismatch obtained for the 7 layer film. The angular separation in the smaller angle stationary points of these curves is approximately 0.12° with the experimental uncertainty being ± 0.04°.

One simple explanation for these observations is that the optical frequency of the experiment coincides with that of a single excitation of the dye multilayers. The single pole model for the dispersion around such an excitation[17], contains two frequency domains where the differential responses of $\varepsilon^{(r)}$ and $\varepsilon^{(i)}$ are in the opposite sense to each other. However, a normal incidence absorption study of this alternated dye structure has shown that the fundamental absorption bands are around 460 nm with negligible absorption at the laser wavelength of 632.8 nm.

An interesting alternative explanation rests on the speculation that the thinner films can undergo field induced structural re-organisations such as orientational re-alignment. This is the mechanism responsible for switching in smectic liquid cystals and there are certain similarities in structure between these and LB films. Recent structural studies made of fatty acid LB films point to a hexatic smectic structure[18,19] and the evidence we have from second harmonic generation[5] and electron diffraction[20] studies indicates that hemicyanine films are composed of tilted chromophores.

Unlike films of ω-tricosenoic acid[21] the macroscopic orientational structure of hemicyanine films suggests no preferential orientation in the dipping direction. This has important consequences for the measurement of the effective loss in these films using surface plasmon spectroscopy.

In the ATR studies, the beam diameter of ~1 mm is much greater than the secondary structure (crystallite regions) and thus probes a range of dielectric tensors. Our measurements of $\varepsilon^{(r)}$ therefore represent some average value. $\varepsilon^{(i)}$ is a measure of the broadening in the ATR minimum and as such includes the effect of scattering due to small scale inhomogeneities and larger scale fluctuations in coupling angle. These latter features of the film structure which one of us has termed 'macroscopic orientational inhomogeneities', (MOI), may account for up to 25% of the measured effective loss[22]. The film re-organisations that we propose here would therefore modify, not only the real part of the dielectric constant but also $\varepsilon^{(i)}$.

If we extend this postulation to describe the features of the thicker film responses, it becomes clear that structural re-organisations are negligible. Since the magnitude of the response for these films depends on the measured dielectric constant changes which arise from all sources, it is plausible that the markedly low response from the 9 layer film signals the removal of one or more of these sources.

4. CONCLUSIONS

We have shown that the amplitude modulation of the reflected beam in the ATR electro-optic configuration varies in a non-simple way with increasing LB layer thickness. We can distinguish two film thickness regimes where the total contributions to the electro-optic response are noticeably different. The thin films may behave similarly to smectic liquid crystals which allow field induced re-alignment of the molecules. On this model in the thick film limit, we suggest that molecular re-alignment is hindered. Experimentally, it should prove possible to de-convolve re-alignment effects from purely electronic polarisations, by performing these measurements at high frequencies.

In addition to reinforcing our view that this technique is of value in determining the magnitude, sign and phase of the electro-optic nonlinear coefficient we have shown that additional structural information may be gained.

5. ACKNOWLEDGEMENTS

This work was supported in part by the Department of Trade and Industry, JOERS collaborative scheme. We thank Neil A Cade for advice on the theoretical treatment of these results.

6. REFERENCES

1. D.J. Williams, "Organic polymeric and non-polymeric materials with large optical nonlinearities", Angew. Chem. Int. Ed. (Engl.) 23, 690-703, (1984)

2. B.F. Levine, C.G. Bethea, C.D. Thurmond, R.T. Lynch and J.L. Bernstein, "An organic crystal with an exceptionally large optical second harmonic coefficent: 2-methyl-4-nitroaniline", J. Appl. Phys 50, 2523-2527, (1979)

3. O.A. Aktsipetrov, N.N. Akhmediev, E.D. Mishina and V.R. Novak, "Second-Harmonic generation on reflection from a monomolecular Langmuir layer", JETP Lett. 37, 207-209, (1983)

4. I.R. Girling, P. V. Kolinsky, N.A. Cade, J.D. Earls and I.R. Peterson, "Second harmonic generation from alternating Langmuir-Blodgett films", Opt. Commun. 55, 289-292, (1985)

5. I.R. Girling, N.A. Cade, P.V. Kolinsky, J.D. Earls, G.H. Cross and I.R. Peterson, "Observation of second harmonic generation from Langmuir-Blodgett multilayers of a hemicyanine dye", Thin Solid Films, 132, 101-112, (1985)

6. D.B. Neal, M.C. Petty, G.G. Roberts, M.M. Ahmad, W.J. Feast, I.R. Girling, N.A. Cade, P.V. Kolinsky and I.R. Peterson, "Second harmonic generation from LB superlattices containing two active components", Electron. Lett. 22, 460-462, (1986)

7. G.H. Cross, I.R. Girling, I.R. Peterson and N.A. Cade, "Linear pockels response of a monolayer Langmuir-Blodgett film", Electron. Lett. 22, 1111-1113, (1986)

8. G.H. Cross, I.R. Girling, I.R. Peterson, N.A. Cade and J.D. Earls, "Optically nonlinear Langmuir-Blodgett films: linear electro-optic properties of monolayers", J. Opt. Soc. Am B, 4, 962-967, (1987)

9. I.R. Girling, N.A. Cade, P.V. Kolinsky, G.H. Cross and I.R. Peterson "Surface plasmon enhanced SHG from a hemicyanine monolayer", J. Phys. D. 19, 2065-2075, (1986)

10. Y.J. Chen and G.M. Carter, "Measurement of third order nonlinear susceptibilities by surface plasmons", Appl. Phys. Lett., 41, 307-309 (1982)

11. I. Pockrand, J.D. Swalen, J.G. Gordon II and M.R. Philpott, "Surface plasmon spectroscopy of organic monolayer assemblies", Surf. Sci. 74, 237-244, (1977)

12. I. Pockrand, J.D. Swalen, R. Santo, A. Brillante and M.R. Philpott, "Optical properties of organic dye monolayers by surface plasmon spectroscopy", J. Chem. Phys. 69, 4001, (1978)

13. A. Brillante and I. Pockrand, "Experimental study of exciton-surface plasmon interactions", J. Mol. Struct. 79, 169-172, (1982)

14. G.M. Wysin, H.J. Simon and R.T. Deck, "Optical bistability with surface plasmons", Opt. Lett. 6, 30-32, 1981)

15. B. Bosacchi and L.M. Narducci, "Optical bistability in the frustrated-total-reflection optical cavity", Opt. Lett. 8, 324-326, (1983)

16. P. Martinot, S. Laval and A. Kostev, "Optical bistability from surface plasmon excitation through a nonlinear medium", J. Phys. (Fr.), 45, 597-600, (1984)

17. G.H. Cross, N.A. Cade, I.R. Girling, I.R. Peterson and D.C. Andrews, "Optically nonlinear Langmuir-Blodgett films: influence of the substrate on film structure and linear optical properties", J. Chem. Phys. 86, 1061-1065, (1987)

18. I.R. Peterson and G.J. Russell, "The deposition and structure of Langmuir-Blodgett films of long-chain acids", Thin Solid Films, 134, 143-152, (1985)

19. I.R. Peterson, "A structural study of the conducting defects in fatty acid Langmuir-Blodgett monolayers", J. Mol. Electron. 2, 95-99, (1986)

20. J.D. Earls, I.R. Peterson, G.J. Russell, I.R. Girling and N.A. Cade, "An electron diffraction study of optically non-linear hemicyanine LB films", J. Mol. Electron 2, 85-94, (1986)

21. W.L. Barnes and J.R. Sambles, "Guided optical waves in Langmuir-Blodgett films of 22-tricosenoic acid", Surf. Sci. 177, 399-416 (1986)

22. I.R. Peterson, J.D. Earls and I.R. Girling, "Orientational inhomogeneity and scattering in Langmuir-Blodgett films of 22-tricosenoic acid", Submitted to Thin Solid Films (1987)

Exceptional Second-order Nonlinear Optical Susceptibilities in Organic Compounds

H.E. Katz, C.W. Dirk

AT&T Bell Laboratories, 600 Mountain Avenue, Murray Hill, New Jersey 07974

K.D. Singer, J.E. Sohn

AT&T Engineering Research Center, P.O.Box 900, Princeton, New Jersey 08540

Abstract

Molecular second-order nonlinear optical susceptibilities (β) of donor-acceptor substituted organic compounds were measured in dimethyl sulfoxide solution using electric field induced second harmonic generation. The experimental data and theoretical interpretations of the trends led to compounds with values of β useful for imparting high bulk susceptibilities on polymeric materials. Exceptionally high values of β were reached relative to analogues of similar size through the use of tricyanovinyl and dithiolylidenemethyl substituents.

Introduction

Organic nonlinear optical materials are expected to be of value in the fabrication of guided-wave optical devices.[1] In order to utilize second-order effects, it is necessary to design and prepare stable, highly optically nonlinear materials that can be processed into light-guides of good optical quality. Glassy polymers offer several inherent advantages[2] over other classes of matter in terms of processability and optical transmission. These polymers are transformed into nonlinearly active ones by the incorporation of molecules or subunits possessing high values of the second-order coefficient β and orienting them in an electric field, as demonstrated[3] for solutions of Disperse Red 1 (DR1)

DR1

in poly(methyl methacrylate). Although this system exhibited the bulk second-order polarizability $\chi^{(2)}$ that was predicted from the β value of DR1 and a thermodynamic model of the poled solution, significant improvements are necessary to obtain a practical device material. Such improvements would be realized through augmentation of the molecular hyperpolarizabilities and dipole moments, and by increasing the density, net orientation, and directional stability of the active species in the polymer matrix. This presentation focuses on optimization of relevant molecular properties using synthetic and physical organic chemical methodology.

It is well known that the generic organic molecule useful in second-order nonlinear optics consists of a strongly electron donating group linked to an electron acceptor by some conjugated bridge.[4] Many of the compounds evaluated to date have been variations on a theme of nitro acceptors and amino or oxy donors.[5] Also, measurements of β have not generally been performed under consistent conditions for different molecules. Our aim is to study molecules with substituents that are strong electron donors or acceptors, and directly measure their second-order susceptibilities in a manner that permits systematic conclusions to be drawn.

Results

The compounds studied are listed in Table 1. Some of the more interesting compounds required new synthetic sequences; two examples are presented in Scheme 1. Note the extensive use of protecting groups to insure selective reactions and the importance of Wittig condensations in assembling the molecular components. Such cyanovinyl acceptors were suggested by Stamatoff et al. for their potential use in nonlinear optical materials.[6]

Experimental values of $\beta\mu$ were obtained for a series of organic compounds as solutions in dimethyl sulfoxide (DMSO) using electric field induced second harmonic generation (EFISH).[7] This technique directly measures the relevant nonlinear optical properties of the molecule. The use of aprotic solvents, such as DMSO and dioxane, is critical to any measurement of β, since deducing molecular properties from solution measurements requires the use of macroscopic local field models which break down for the short-range interactions and associations present in protic solvents such as methanol. The macroscopic nonlinear optical susceptibility of a poled polymer glass incorporating a nonlinear optical moiety is directly proportional to $\beta\mu$, which is the quantity measured using EFISH techniques.[3] Values of β may be obtained by dividing by dipole moment magnitudes. A comparison of the contributions of molecular moments to β is made through the quantity β_0, in which the dependence of β on the electronic transition energies of the compounds is approximately factored out using a two level model which is given by,[8]

$$\beta_{xxx}(-2\omega;\omega,\omega) = \frac{e^3|\mu_{01}|^2(\mu_{11}-\mu_{00})}{h^2} \times F(\omega) \qquad (1)$$

where

$$F(\omega) = \frac{3\omega_0^2}{(\omega_0^2-\omega^2)(\omega_0^2-4\omega^2)} \qquad (2)$$

and where e is the electronic charge, the μ's are molecular moments. The quantity β_0 is calculated by dividing the measured value of β by $F(\omega)$ at the measured fundamental frequency, and multiplying by $F(0)$. The data are obtained at a fundamental wavelength of 1.356 μm, and are summarized in Table 1.

For a series of three molecules (1-3) differing only in the acceptor group, there is a dramatic increase in β and β_0 in the sequence nitro, dicyanovinyl, tricyanovinyl. A correlation is obtained on plotting β_0, a resonance-dominated quantity, vs σ_R^-, a resonance-dominated Hammett constant (Figure 1), suggesting that β increases with ground state resonance.

A more sophisticated approach is to employ semi-empirical molecular orbital calculations of π-electron configuration[9] in order to theoretically predict the trends in β and μ for this series of compounds. Figure 2 is a dispersion plot of theoretical β as a function of the second harmonic energy of the probe beam. This plot emphasizes differences due to dispersion, since all three curves rise sharply near the transition frequencies of the respective molecules and shows that β increases as the second harmonic of the measurement frequency approaches the first excitation energy. Figure 3 shows the same three curves positioned relative to an arbitrary common transition frequency ω_1. Here the ordering is due to differences in the contribution of molecular moments to the hyperpolarizabilities. The tricyanovinyl group induces greater changes in dipole moment upon excitation, and also leads to a much better projection of the ground state dipole moment on the main hyperpolarizability component relative to the dicyanovinyl substituent.

We also examined a series in which the donor was varied and the acceptor was kept constant (1,4-6). The quantity $\beta\mu$ changes very little when the amino group of p-nitroaniline is modified, but increases substantially when it is replaced by the easily ionized dithiolylidinemethyl group (6). A value for σ_p^+ for this dithiole donor was obtained from ^{13}C NMR, and a Hammett plot of β_0 vs σ_p^+ is shown in Figure 4; no correlation is observed. The main contributor to increased β_0 in the dithiole compound is an excited state-related moment, since the ground state Hammett constants do not predict such an increase, as they did in the case of the acceptor sequence.

By assembling the best electron donor and acceptor in this series (8), the highest value of $\beta\mu$ for a disubstituted benzene is achieved. Additionally, the generally accepted hypothesis that longer conjugated π-electron systems give rise to larger values of β and $\beta\mu$ within series in which both donor and acceptor are constant (for example, 1 and DR1, 2 and 7), is confirmed. By incorporating the tricyanovinyl acceptor in an amino-substituted azo dye (9), an unusually high value of β is obtained that is several times larger than that of the azo dye DR1, without substantially increasing the size of the chromophore.

Conclusion

We have studied several structural variables relevant to hyperpolarizabilities in organic molecules using infinite dilution solution measurements of electric field induced second harmonic generation. Using this systematic and direct measurement of β on a diverse series of compounds, molecular engineering techniques have been extended to produce species possessing exceptional second-order nonlinear optical susceptibilities.

Table 1

MOLECULE	$\beta\mu$	β	β_0
1	138	21	12
2	271	31	16
3	846	78	26
4	75#	12	9
5	102	15	9.5
6	358	52	25
7	2650	323	133
8	1200#	–	–
9	4110#	390	154
DR1	1090	125	47

$\beta\mu$ in units of $10^{-30} cm^5 D/esu$; β, β_0 in units of $10^{-30} cm^5/esu$

Measured at $\lambda = 1.58\mu m$

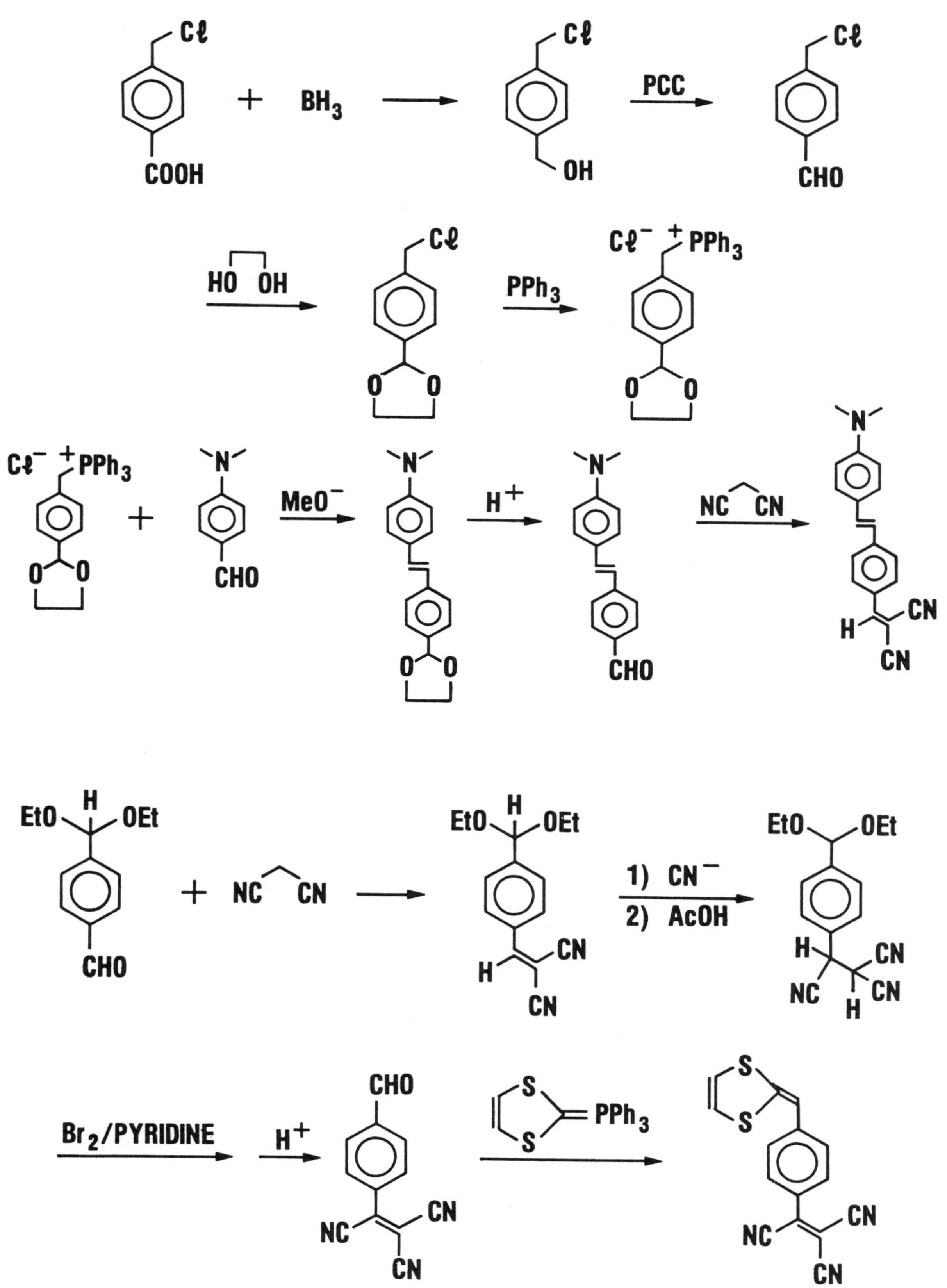

Cℓ
COOH
BH3
Cℓ
OH
PCC
Cℓ
CHO
HO OH
Cℓ
O O
PPh3
Cℓ⁻ ⁺PPh3
O O
Cℓ⁻ ⁺PPh3
O O
N
CHO
MeO⁻
N
O O
H⁺
N
CHO
NC CN
N
H
CN
CN
EtO H OEt
CHO
NC CN
EtO H OEt
H
CN
CN
1) CN⁻
2) AcOH
EtO OEt
H
CN
NC
H
CN
Br2/PYRIDINE
H⁺
CHO
NC
CN
CN
S
S
PPh3
S
S
NC
CN
CN

Scheme 1

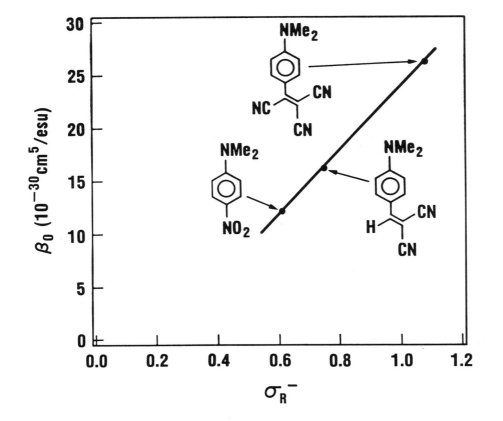

Figure 1

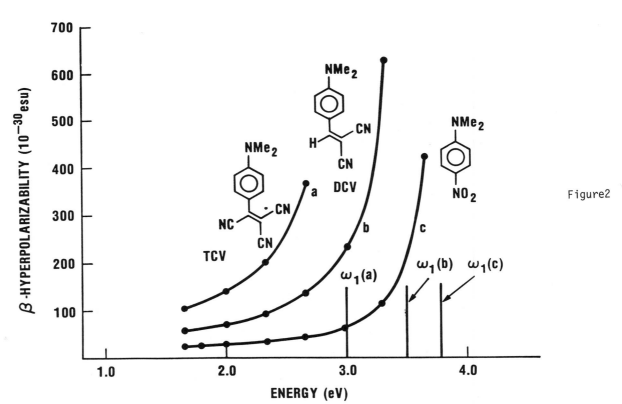

Figure2

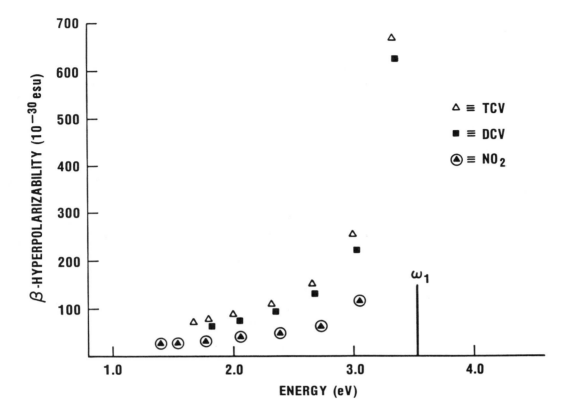

Figure 3

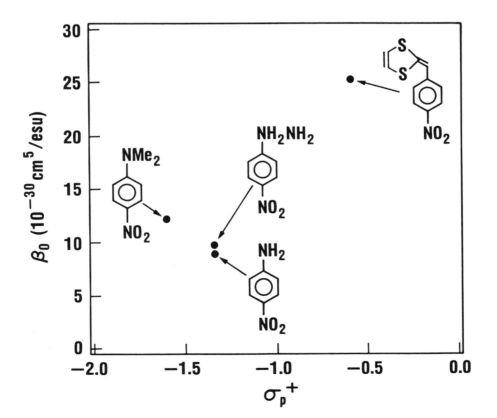

Figure 4

Bibliography

1. Zyss, J. *J. Molec. Electron.* **1985**, *1*, 25.

2. Tomlinson, W.J.; Chandross, E. A. in *Advances in Photochemistry* **1979**, 12.

3. Singer, K.D.; Sohn, J.E.; Lalama, S.J. *Appl. Phys. Lett.* **1986**, *49*, 248.

4. Chemla, D. S.; Zyss, J. eds. *Nonlinear Optical Properties of Organic Molecules and Crystals*: Academic Press, Orlando, 1987.

5. Twieg, R.J. and Jain, K. in Williams, D.J., Ed., *Nonlinear Optical Properties of Organic and Polymeric Materials*, ACS Symp. Ser. No. 233, Washington, D.C. 1983.

6. Stamatoff, J. *et al. Proc. SPIE* **1986**, *682*, 85.

7. Singer, K.D.; Garito, A.F. *J. Chem. Phys.* **1981**, *75*, 3572.

8. Oudar, J.L. *J. Chem Phys.* **1977**, *67*, 446.

9. Dirk, C.W.; Twieg, R.J.; Wagniere, G. *J. Am. Chem. Soc.* **1986**, *108*, 5387.

NEW POLYDIACETYLENES INCORPORATING MESOGENIC SIDE-CHAINS

Michael A. Schen

National Bureau of Standards, Institute for Material Science and Engineering,
Polymers Division, Blg. 224, Rm.B320, Gaithersburg, MD 20899 USA

ABSTRACT

The categorization of existing polydiacetylene structures according to the nature of the polymer side-chain interactions is presented. Opportunities for generating new diacetylene molecular architectures based on creating mesogenic side-chains then becomes evident. Carrying through on this concept, the synthesis of four new symmetric diacetylene monomers based on the N-benzylideneaniline and cyanobiphenyl mesogens is reported. 1,6-Bis[N-(4-oxybenzylidene) 4-octylaniline] 2,4-hexadiyne(1-OBOA) contains a methylene oxide spacer between the mesogen and the diacetylene core and shows smectic liquid crystalline textures over a broad temperature range. Monomers containing a tetrametylene-oxide spacer do not show liquid crystallinity as monomers but are intended to result in liquid crystalline polymer.

1. INTRODUCTION

For the integrated optics engineer, a strong need exists for organic and inorganic materials which may be fabricated into switching, logic or phase conjugating devices. Not only must these materials satisfy or exceed the necessary primary non-linear optical requirements such as second order(χ^2) or third order(χ^3) susceptibility or switching rate(τ), it must simultaneously satisfy a variety of secondary requirements such as high optical quality, material-substrate adhesion, compatible micro and macromorphologies, environmental stability, high laser damage threshold energy, and low dielectric constant to name a few. It is the challenge of the materials scientist to keep into perspective the importance of all of these factors for directing their work in this field.

It has been shown that organic small molecules and polymers containing highly charge correlated π-electron systems represent a class of materials with large intrinsic non-linear susceptibilities(β, γ,...)[1,2,3]. Specifically, polydiacetylenes(PDA's) exhibit exceptionally large non-resonant χ^3(ca. 10^{-10} esu)[4,5], fast switching times(ca. 1 psec)[5], phase matching[2], and optical transparency throughout much of the visible and near infra-red regions. PDA's are therefore very likely candidates for future device applications.

1.1 Existing Diacetylene Structures

A number of symmetrically disubstituted monomers have been synthesized and polymerized consequent to the works of Wegner and coworkers[6]. Some of these appear along with their abbreviations in **Table 1.1**. Focusing on the nature of intermolecular interactions between R groups and the geometric anisotropy of the molecule, four types of monomer structures become evident. In type A monomers, R group interactions are limited strictly to van der Waals forces within the crystal and the melt. In type B monomers, hydrogen bonding between molecules can exist in the crystal, melt, and in solution depending on the choice of solvent. In type C monomers, the possibility for specific intermolecular charge transfer excimer or exciplex formation between neighboring residues exists. All monomer molecules within types A, B, or C are also highly flexible chain structures and simply undergo crystal-isotropic phase transitions upon heating. With polymerization, only a few polymers arising from the above monomer examples show appreciable solubility or exhibit reversible melting. Consequently, polymer macromorphologies are usually limited to individual small crystals. Reprecipitated films from soluble polymer such as n-BCMU may be processed into a variety of formats but contain trapped configurational defects resulting in a red shift in the optical absorption spectra[7].

In contrast, type D monomers are rigid highly anisotropic molecules and exhibit liquid crystalline phase behavior at elevated temperatures. However, the DVDA's[2] are not the only examples of diacetylene based liquid crystals. Symmetric diacetylene liquid crystals of the type shown in **Figure 1.2** have been reported separately by Grant et al.[8] and Ozcayir et al.[9a]. Because no mention has been made however of the response of this family of hindered diacetylene molecules to UV, X-ray, or [60]Co γ-ray irradiation, the question of polymerizability remains unanswered. Polyesters and copolyesters incorporating these groups within the main chain do display however, nonreversible thermal behavior when cycled above 200° and may be due to a degree of chain extension of the diacetylene moieties[9b].

Table 1.1

Symmetric Diacetylene Monomers

$$R - C \equiv C - C \equiv C - R$$

Abbreviation	R	Monomer Type[a]	Polymer Type[b]
TS, PTS	$-CH_2-O-(SO_2)-C_6H_4-p-CH_3$	A	i
TS-12	$-(CH_2)_4-O-(SO_2)-C_6H_4-p-CH_3$	A	s/x-i
BPG	$-C_6H_4-p-COO-(CH_2)_3-OOC-C_6H_5$	A	i
TCDU	$-(CH_2)_4-OOC-NH-C_6H_5$	B	i
ETCD	$-(CH_2)_4OOC-NH-C_2H_5$	B	s
n-BCMU	$-(CH_2)_n-OOC-NH-CH_2-COO-C_4H_9$ $n = 3,4$	B	s/x-i
DCH	$-CH_2-N(C_{12}H_8)$	C	i
DVDA	$-C=CH-C_6H_4-(p)-R'$ $R' = NH_2, OCH_3, OC_4H_8, CN$	D	i

----
[a] A: Van der Waals forces; B: hydrogen bonding; C: charge-transfer interacting; D: liquid crystalline.
[b] i: insoluble; s: soluble; x-i: crystalline-isotropic melting.

$$R \langle\rangle C \equiv C - C \equiv C \langle\rangle R$$

R = n-alkyl[8]
n-alkoxy[8]
n-alkanoyloxy[9a]

Figure 1.2: Previously described diacetylene liquid crystals.

1.2   New Architectures

The polymerization of DVDA has been demonstrated to take place spontaneously upon heating not within the condensed crystal phase but at higher temperatures within the nematic liquid crystalline phase[2]. Inhomogeneous conversion of monomer to polymer ultimately leads to the coalescence of polymer droplets into a continuous matrix. To date, a thorough examination into the chemistry and physics of this system has not been published. This behavior to undergo polymerization in the nematic phase however does point to a potential new approach to the synthesis of polydiacetylenes. Specifically, by starting with macroscopically ordered liquid crystalline monomer, polymer structures of high optical performance and of various geometric formats may be possible. Whether the polymerization chemistry characteristic of diacetylenes in the solid crystal can be affected within the ordered liquid crystalline phase to yield high polymer which is also highly ordered is a critical question which must be addressed.

In this communication, the synthesis of new symmetric diacetylene monomers incorporating the N-benzylideneaniline and cyanobiphenyl mesogen moieties is reported along with a thorough characterization of their optical and thermal behavior. Figure 1.3 shows in block form the three classes of monomer and polymer architectures described. Reactivities towards polymerization are discussed along with macroscopic polymer properties. Some of these materials have been designed with the anticipation of observing monomeric liquid crystalline phase behavior(Fig. 1.3(a)). Subtle variation of the monomer architecture to include a flexible spacer between the diacetylene core and the rigid mesogen(Figs. 1.3(b, c)) creates a structure which is likely to yield polymer that is soluble and thermotropic liquid crystalline. Therefore, these new monomer and polymer architectures introduce a new level of side-chain organization within this family of polymers.

2.   EXPERIMENTAL RESULTS

2.1   Monomer Synthesis

Representative examples of Class 1, 2, and 3 diacetylene mesogenic monomers which have been synthesized are listed in Table 2.1.

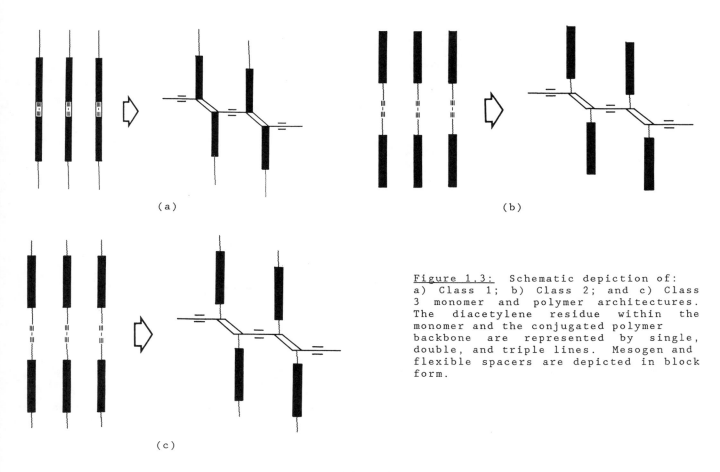

(a)

(b)

(c)

Figure 1.3: Schematic depiction of: a) Class 1; b) Class 2; and c) Class 3 monomer and polymer architectures. The diacetylene residue within the monomer and the conjugated polymer backbone are represented by single, double, and triple lines. Mesogen and flexible spacers are depicted in block form.

Table 2.1

Mesogenic Diacetylene Monomers

Structure	Abbreviation	Class
$\left[C{\equiv}CCH_2O\bigcirc CH{=}N\bigcirc(CH_2)_7CH_3\right]_2$	1-OBOA	1
$\left[C{\equiv}C(CH_2)_4O\bigcirc N{=}CH\bigcirc CN\right]_2$	4-CBOA	2
$\left[C{\equiv}C(CH_2)_4O\bigcirc{-}\bigcirc CN\right]_2$	4-COB	2
$\left[C{\equiv}C(CH_2)_4O\bigcirc CH{=}N\bigcirc(CH_2)_7CH_3\right]_2$	4-OBOA	3

Scheme 2.1 exemplifies the synthesis of 1-OBOA by classical means. Similarly, scheme 2.2 was found more appropriate for 4-OBOA, 4-CBOA, and 4-COB. Final products were purified by repeated extractions to remove unreacted starting materials followed by multiple recrystallization or liquid chromatography. FTIR, NMR, and liquid chromatography showed the final monomers to be of high purity. A Zeiss microscope equipped with cross polarization and 1λ retardation filters, Zeiss MC63/M35 photographic system and Metler PF2 hot stage were used for optical characterization. Optical scans rates were typically 2 or $10°min^{-1}$. Differential scanning calorimetry(DSC) and thermal gravimetric analysis(TGA) data were collected using Perkin-Elmer DSC2 and DSC7/TGA7 workstations at 20 and 10 deg./min. scan rates[10]. A 100 watt medium pressure mercury lamp with emissions from 185-366 nm was used for UV irradiation experiments.

Scheme 2.1

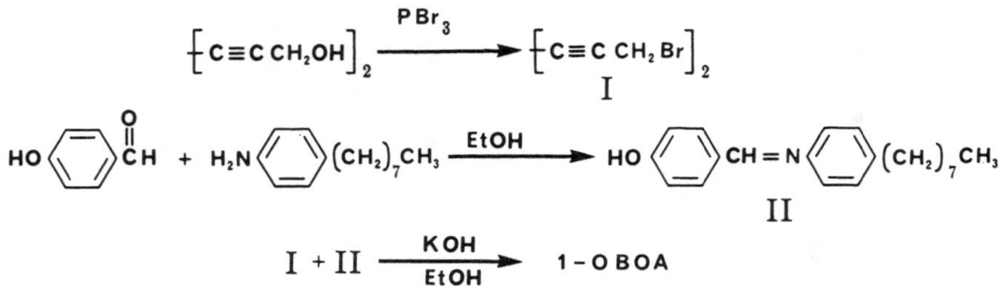

$$\left[C \equiv C\,CH_2OH\right]_2 \xrightarrow{PBr_3} \left[C \equiv C\,CH_2Br\right]_2 \quad \text{I}$$

$$HO\text{-}\phi\text{-}\overset{O}{\underset{}{C}}H \;+\; H_2N\text{-}\phi\text{-}(CH_2)_7CH_3 \xrightarrow{EtOH} HO\text{-}\phi\text{-}CH=N\text{-}\phi\text{-}(CH_2)_7CH_3 \quad \text{II}$$

$$\text{I} + \text{II} \xrightarrow[EtOH]{KOH} 1\text{-}OBOA$$

Scheme 2.2

$$HC \equiv C(CH_2)_4OH \xrightarrow[TMEDA]{CuCl} \left[C \equiv C(CH_2)_4OH\right]_2 \xrightarrow{PBr_3} \left[C \equiv C(CH_2)_4Br\right]_2 \quad \text{I}$$

$$\text{I} + 2\,HO\text{-}\phi\text{-}\overset{O}{\underset{}{C}}H \xrightarrow[KOH]{EtOH} \left[C \equiv C(CH_2)_4O\text{-}\phi\text{-}\overset{O}{\underset{}{C}}H\right]_2 \xrightarrow[EtOH]{2[H_2N\,\phi\,(CH_2)_7CH_3]} 4\text{-}OBOA$$

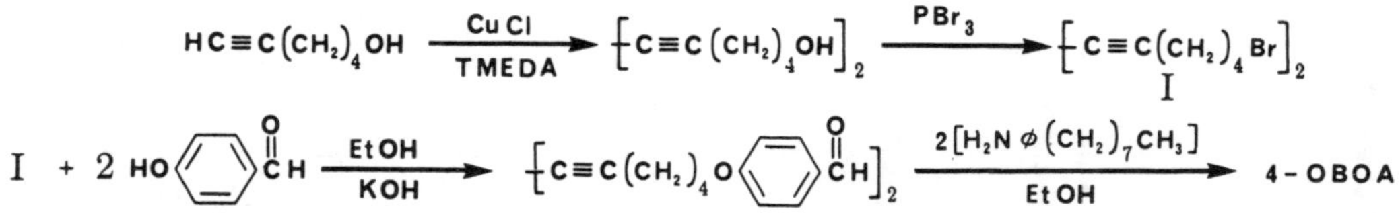

2.2 Monomer Characterization.

Room temperature optical microscopy showed 4-COB and 4-OBOA to exist in the form of needle-like crystals of colorless and creamy-yellow colors respectively. Independent of recrystallization solvent, 4-CBOA could only be recovered as small plate-like crystals of light-yellow color. Monomer 1-OBOA could never be recovered with any defined crystal morphology. Instead, a highly birefringent irregularly shaped waxy precipitate of light tan color resulted. Birefringent thin films of 1-OBOA could also be obtained by shearing the material across the microscope slide at room temperature. This behavior is similar to that of plastic crystals though the molecular geometries of the two are vastly different.

TGA data was collected on 1-OBOA, 4-COB and 4-OBOA under air and nitrogen atmospheres to determine the maximum operating temperature before which degradative weight loss is observed. A 1% loss in mass was the criterion for defining this limit. Initial weight losses at 235°, 255°, and 318°C for 4-COB, 1-OBOA, and 4-OBOA respectively were recorded.

DSC heating of 4-CBOA showed a single, sharp melting transition at 158°C upon heating and a recrystallization exotherm at 114°C. Optically, a crystal-to-isotropic phase transition was seen to begin at 158°C and was completed by 167°C at $2° \cdot min^{-1}$ heating. Therefore 4-CBOA monomer does not show liquid crystalline behavior upon heating.

The DSC of 4-COB showed however a more interesting behavior. Upon initial heating, an endothermic maxima and shoulder at 86° and 77°C respectively were observed. No exothermic recrystallization was observed on cooling even at rates as slow as $1.25° \cdot min^{-1}$. Subsequent reheating resulted in extensive recrystallization at 39°C followed by melting at 88°C. The shoulder at 77°C has now been further resolved into a second peak at 78°C. No deviation from this behavior was observed with repeated cycling.

Optical microscopy of 4-COB showed no signs of melting until 84°C with the sample going totally isotropic at 96°C. Therefore, 4-COB also does not exhibit liquid crystalline behavior. Cooling from the isotropic melt to room temperature did not result in recrystallization but annealing at 35°C did show very slow formation of crystal nuclei. Self seeding recrystallization starting from the nearly totally melted material resulted in extensive recrystallization by 69°C giving oblong shaped crystals. These results indicate a very high barrier for the formation of critical sized nuclei needed for crystallization. During cooling from the totally isotropic melt, recrystallization proceeds very slowly once nuclei are formed presumably due to the high melt viscosity. Consequently, recrystallization occurs only on heating as the viscosity drops. The existence of two endothermic transitions on heating may also arise from crystalline polymorphism. X-ray analysis would be needed however to verify this hypothesis.

The possibility of crystalline polymorphism may also explain the DSC results for 4-OBOA. Two sharp endothermic transitions at 84° and 124°C were observed with enthalpies of 5.37 and 14.19 $kcal \cdot mol^{-1}$ respectively. Optical examination also revealed only a singular crystal-to-isotropic transition at 123°C with no liquid crystalline behavior observed.

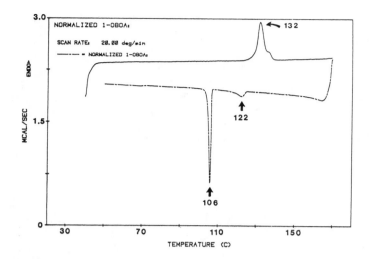

Figure 2.1: DSC heating and cooling scans of fresh 1-OBOA with no high temperature isothermal annealing between scans.

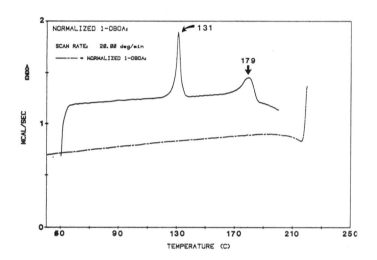

Figure 2.2: Second DSC heating and cooling scans of 1-OBOA with isothermal annealing near 190°C.

As was earlier mentioned, 1-OBOA exists at room temperature as a crystalline waxy material with a very low shear modulus. **Figure 2.1** shows the DSC trace from 40° to 170°C with an endothermic maximum and shoulder at 132° and 136°C respectively. No deviation from flat baseline is observed between 40° and -70°C also. Cooling from 170°C shows two exothermic transitions at 122° and 106°C. If heating beyond 170°C is allowed however(**Figure 2.2**), a second endothermic transition is observed at 179°C with baseline deviations observed there-after. If held at temperatures near this second transition, subsequent cooling shows no recrystallization and reheating curves are equally featureless.

A series of optical micrographs representative of this behavior is shown in **Figures 2.3(a-e)**. Below 132°C, a poly-crystalline semi-rigid solid is seen as shown in Fig. 2.3(a). Shear lines remaining from the initial spreading of the material can be seen. Fig. 2.3(b) was taken at 145°C of the same region showing liquid crystalline birefringence above the 132°C first order transition. However, though the sample has undergone a crystal-to-liquid crystal transition, no long range flow has occurred. This indicates a non-zero shear modulus within this mesophase. In fact, sample flow is not observed until ca. 170°C. Loss of birefringence occurs shortly thereafter at 179°C. Subsequent cooling from the isotropic melt leads to an abrupt reappearance of liquid crystallinity at 169°C with a well defined schlieren texture observed by 160°C(Fig. 2.3(c)). Cooling further to 121°C abruptly gives rise to a mosaic texture(Fig. 2.3(d)) with domains of 10-20ηm dimensions. If 1-OBOA is held isothermally at ca. 150°C or above, microphase separation of orange colored, non-birefringent microdomains develops which then coalesce over time to form large droplets(Fig. 2.3(e)). The entire sample eventually forms this second phase which upon cooling does not recrystallize.

Figure 2.4(a) and (b) show the infra-red spectra of 1-OBOA before and after annealing at 150°C respectively. As can be seen, no change in the spectra has occurred therefore ruling out degradation as a cause of this phase change.

2.3 Monomer Reactivities.

Limited information has been gathered so far on the polymers arising from the above described monomers. However, experiments have identified the sensitivities of these compounds to polymerization. All compounds with the exception of 1-OBOA are thermally stable through their melting transitions giving roughly constant DSC heats of fusion over multiple heating and cooling cycles. As shown earlier, thermal annealing of 1-OBOA at or above 150°C irreversibly results in a phase change. Because color formation is associated with the build up of pi-conjugation in diacetylene polymerizations, the generation of a dark orange phase upon heating of 1-OBOA suggests some degree of chain extension. This behavior is similar to that of the DVDA's reported by Garito[2].

Sensitivity to UV irradiation was found to decrease in the following order: 4-COB > 4-OBOA > 4-CBOA ≃ 1-OBOA. Poly4-COB appears as deeply burgundy colored metallic looking crystals and poly4-OBOA as dark orange crystals of a curled ribbon-like macro-morphology. 4-CBOA and 1-OBOA are essentially insensitive to UV irradiation.

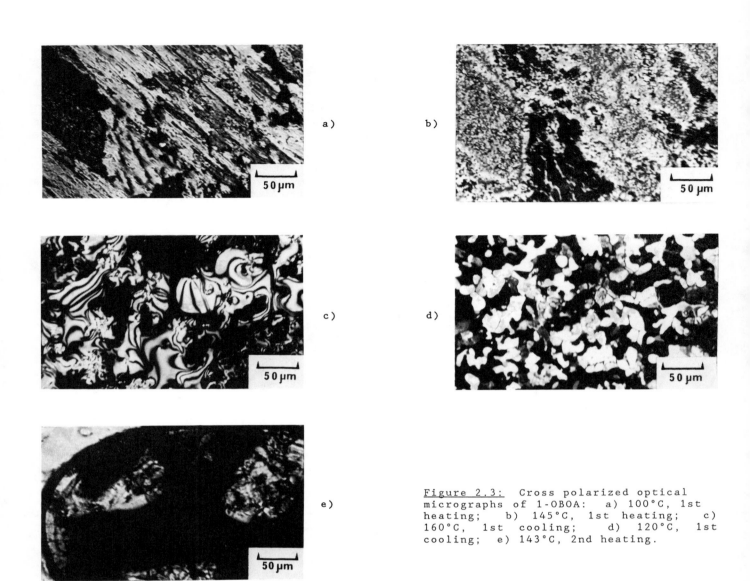

Figure 2.3: Cross polarized optical micrographs of 1-OBOA: a) 100°C, 1st heating; b) 145°C, 1st heating; c) 160°C, 1st cooling; d) 120°C, 1st cooling; e) 143°C, 2nd heating.

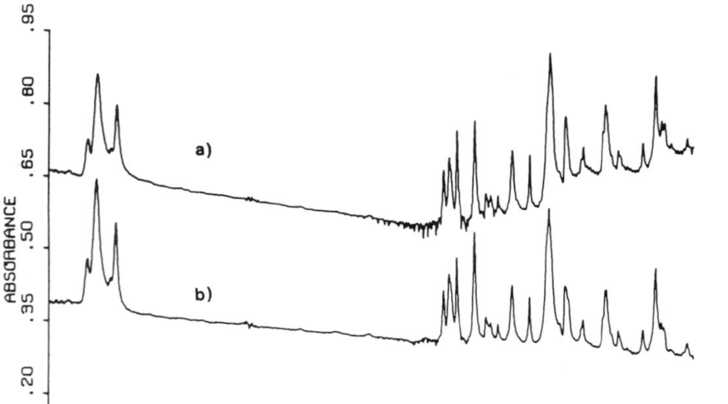

Figure 2.4: FTIR spectra of 1-OBOA: a) nascent; b) after 150°C, 240 min. annealing.

3.0 DISCUSSION

The observation that Class 2 and 3 monomers(**Table 2.1**) do not exhibit liquid crystalline phase behavior on heating is not very surprising. With a flexible tetramethylene-oxide 'soft' segment bridged between the 'rigid' benzylidene mesogens and the 'rigid' diacetylene core, a variety of molecular conformations are possible. This 'blocky' architecture is characteristic of many thermotropic main-chain liquid crystalline polymers, but in these cases, mesophase formation is seen only with polymer of high molecular weight. The suspected crystal phase polymorphism in 4-COB and 4-OBOA and the slow nucleation kinetics of 4-COB may also arise from the 'blocky' nature of these monomers. However, after conversion to polymer, these same 'flexible' polymethylene-oxide spacers can provide the necessary conditions for observing polymer liquid crystallinity. If the thermally induced conformational fluctuations of the 'rigid' side-chain can be adequately decoupled from those of the polymer backbone, liquid crystallinity in the melt will be observed.

Introduction of a single methylene-oxide spacer between the diacetylene core and the benzylidene moiety on the other hand, is insufficient to decouple the individual motions of these units and consequently a single rigid unit results(**Class 1**). Liquid crystallinity is consequently observed. From the mesophase textures exhibited by 1-OBOA at various temperatures, tentative microstructures may be assigned based on the observed textures of existing liquid crystals. However, for definitive phase identification, X-ray scattering would be necessary. On cooling the isotropic melt of 1-OBOA, a schlieren texture characteristic of either nematic or smectic C mesophases is observed at 169-170°C. However the threaded nature of the observed pattern makes this more suspect of the smectic C phase. From 169-121°C, roughly the same texture is retained, but then undergoes a dramatic volume shrinkage and mosaic texture formation at 121°C. The volume contraction indicates the development of long range intermolecular packing within a smectic layer. This phenomenon is characteristic of the transition from a smectic A or C phase to a smectic B, F, G, E or H phase. In smectic A and C phases, there is no long range correlation between molecules within a layer. In the smectic B, F and G phases, a hexagonally close packed array of molecules within layers exists. In smectic G structures, correlations between layers is also present. In the E and H phases, the packing is monoclinic. Because large highly colored mosaic textures and a non-zero shear modulus at temperatures just above the crystal-to-smectic phase transition are observed, the smectic G structure is believed likely in 1-OBOA above the crystal-to-smectic phase transition.

The observed formation of a secondary, highly colored phase in thermally annealed 1-OBOA is consistent with observations made on DVDA in the nematic state. The lack of crystallinity and therefore molecular organization within this secondary phase may be a consequence of phase transformation occurring near the clearing temperature of the monomer. In this temperature regime, the monomer is in its most highly disorganized liquid crystalline state. Thermal annealing within the low temperature smectic phase however, does not appear to cause secondary phase formation.

And lastly, the enhanced UV stability of the benzylidene based monomers over 4-COB is probably due to the UV absorptivity of aromatic imines. Consequently, these monomers may be safely handled under ambient lighting without unwanted polymerization. It also indicates that gamma irradiation will be necessary for obtaining high polymer.

REFERENCES

1. G.P. Agrawal, C. Cojan, C. Flytzanis, Phys. Rev. B, 17(2), 776 (1978).

2. A.F.Garito, C.C. Teng, K.Y. Wong, O. Zammani-khamiri, Mol. Cryst. Liq. Cryst., 106, 219, (1984).

3. L.R. Dalton, J. Thomson, H.S. Nalwa, Polymer, 28, 543, (1987).

4. C. Sauteret, J.P. Hermann, R. Frey, F. Pradere, J. Ducuing, R.H. Baughman, R.R. Chance, Phys. Rev. Lettr., 36(16), 956, (1976).

5. P.P. Ho, R. Dorsinville, N.L. Yang, G. Odian, B. Eichmann, J. Jimbo, O.Z. Wang, G.C. Tang, N.D. Chen, W.K. Zou, Y. Li, R.R. Alfano, Proc. SPIE, 682, 36, (1986).

6. G.Wegner, Z. Naturforschg., 24b, 824, (1969).

7. R.R. Chance, G.N. Patel, J.D. Witt, J. Chem. Phys., 71(1), 206, (1979).

8. B. Grant, Mol. Cryst. Liq. Cryst., 48, 175, (1978); E. M. Barrall, B. Grant, A.R. Gregges in Liquid Crystals and Ordered Fluids, J.F. Johnson, R.S. Porter, eds., Vol 3, pg. 19, 1973.

9. a) Y. Ozcayir, J. Asrar, A. Blumstein, Mol. Cryst. Liq. Cryst., 110, 262, (1984); b) Y. Ozcayir, A. Blumstein, J. Polym. Sci., Polym. Chem. Ed., 24, 1217, (1986).

10. Commercial equipment, instruments, or materials are identified in this paper to adequately specify the experimental procedure. Such identification does not imply recommendation or endorsement by the National Bureau of Standards.

ADVANCES IN NONLINEAR POLYMERS AND INORGANIC CRYSTALS,
LIQUID CRYSTALS, AND LASER MEDIA

Volume 824

Session 3

Molecular and Polymeric Optoelectronic Materials II

Chair
Garo Khanarian
Hoechst-Celanese Corporation

Invited Paper

Conjugated polymers for nonlinear optics

Gregory L. Baker, and S. Etemad

Bell Communications Research, Red Bank NJ 07701

F. Kajzar

CEA-IRDA, DEIN/LERA CEN/SACLAY, 91191 Gif-sur-Yvette, France

ABSTRACT

The large nonresonant third order optical nonlinearities and fast response times of conjugated polymers make them attractive materials for nonlinear optics. The $\chi^{(3)}$ spectrum of polyacetylene and polydiacetylenes show similar structure, with magnitudes that agree with calculations of $\chi^{(3)}$ for the infinite polyene, as well as from the extrapolation of the calculated hyperpolarizabilities of finite polyenes. Of the conjugated polymers, the polydiacetylenes offer the best combination of physical properties since they can be obtained as large single crystals, thin crystalline films, Langmuir-Blodgett multilayers or as spun thin films. Unfortunately, all the conjugated polymers show some sensitivity to oxygen, though single crystals of polydiacetylenes are robust.

1. INTRODUCTION

Conjugated organic polymers are now recognized as leading candidates for applications in nonlinear optics [1] Members of this class of polymers contain 1-D planar arrays of alternating single (σ) and double ($\sigma + \pi$) bonds and include the polyacetylenes, polydiacetylenes, cumulenes, and polyynes (Figure 1). Since the cumulenes and polyynes are difficult synthetic targets, most interest has focused on polyacetylene (R = H) and the many varieties of polydiacetylenes.

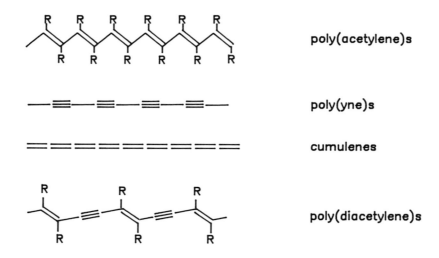

poly(acetylene)s

poly(yne)s

cumulenes

poly(diacetylene)s

Figure 1. The family of conjugated polymers.

Polyacetylenes and polydiacetylenes are attractive materials since they combine large third order nonlinearities with fast response times, suggesting the possibility of low power, ultrafast optical switching. Practical nonlinear materials for such applications must possess more than just a high optical nonlinearity. Useful materials must also have acceptable mechanical properties that enable the fabrication of high quality crystals or defect-free thin films. These films must be of optical quality with little or no scattering of light. The nonlinear material must also be stable under use conditions, especially toward degradation by oxygen. In this contribution, we will look at the spectrum of $\chi^{(3)}$ for the infinite polyene and for the infinite polydiacetylene. We will also briefly examine the issues of the tractability and stability of conjugated polymers.

2. ELECTRONIC PROPERTIES

The motivation for investigating the third order susceptibilities of conjugated polymers results from early work on finite polyenes. [2] The measured hyperpolarizabilities, γ_{xxxx}, of a variety of conjugated hydrocarbons in the off-resonance regime showed that γ_{xxxx} was a strong function of the number of π bonds in conjugation, n. The obvious question is what is the maximum $\chi^{(3)}$ that could be obtained from a polyene system? For the case of the infinite polyene, Agrawal, Cojan, and Flyntzaniz pointed out that the magnitude of the hyperpolarizability should scale with the sixth power of the delocalization length, N_d. [3] Their expression for the third order nonlinearity, $\chi^{(3)}$, of the infinite polyene is:

$$\chi^{(3)} = \chi_\pi^{(3)} = \frac{16}{45\pi}\chi_\sigma^{(3)}N_d^6 = \sigma\frac{e^4a^3\sigma}{90\pi W^3} \times \left[\frac{W}{E_g}\right]^6$$

where e is the electronic charge, a is the unit cell, σ is the cross-section per chain, W is the bandwidth and E_g is the band gap. When parameters appropriate to polyacetylene are used, a value of $\chi^{(3)} = 6.5 \times 10^{-10}$ esu is obtained.

From experiments, it is well known that the optical band gap of finite polyenes scales linearly with $1/n$ where n is the number of double bonds. This relationship holds well for when $n > 4$. Since E_g appears in the denominator of the expression for $\chi^{(3)}$, γ_{xxxx} for a series of polyenes should initially increase rapidly with n, then approach an asymptotic limit. The limited availability of an extended series of isostructural polyenes prevents experimental verification of this trend, but recent calculations of the hyperpolarizabilities of finite polyenes, cumulenes, and polyynes confirms the expectation. [4] The data of Beratan, Onuchic, and Perry has been replotted in Figure 2 to show that log γ_{xxxx} scales as $1/n$. From this plot, a simple linear extrapolation gives the hyperpolarizability of the infinite polyene. The result agrees with the value obtained from the calculations of Agrawal, Cojan and Flyntzaniz, which is shown as the horizontal dotted line in Figure 2. This $1/n$ relationship should also hold for other conjugated polymers. The optical band gaps of polydiacetylene oligomers have a $1/n$ dependency, [5] thus

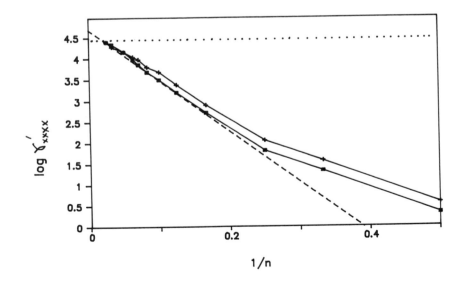

Figure 2. Dependence of the normalized hyperpolarizability density $\gamma'_{xxxx} \equiv \gamma_{xxxx}/N\gamma^0_{xxxx}$ of polyenes (■) and polyynes (+) with chain length. The data is from reference 4 and has been replotted to show the $1/n$ relationship.

the hyperpolarizabilities of polydiacetylenes should exhibit behavior analogous to the polyenes.

Of practical importance is that for a polyene segment of moderate length (n = 10), the hyperpolarizability is within an order of magnitude of the ultimate value. Since the tractability of finite polyenes is superior to the nearly intractable polyacetylene, compromises that sacrifice some of the magnitude of $\chi^{(3)}$ in return for enhanced processability may lead to useful materials.

In Figure 3 we compare the experimental results for the nonresonant $\chi^{(3)}$ of the infinite polyene and polydiacetylene obtained from frequency tripling experiments. Values of $\chi^{(3)}$ have been obtained as a function of ω spanning energies just below the optical gap to well into the infrared. The results for *trans*-polyacetylene are reproduced from our earlier work. [6] The results for PTS have been obtained using thin (a few microns) single crystals grown by the solution-shear method. [7] Since PTS samples of this thickness absorb an appreciable amount of the third harmonic generated in the experiment, the data are valid for qualitative comparisons with that of polyacetylene. The results are similar for both polymers; $\chi^{(3)}$ rises as the energy decreases, with a sharp spike at 0.9 eV. The $\chi^{(3)}$ spectrum is similar to that reported

for Langmuir-Blodgett multilayers. [8] The peak near 0.9 eV has been attributed to the presence of a two-photon accessible state, while the increase in $\chi^{(3)}$ with decreases in energy correspond to three-photon accessible states. The shape of the curve should thus roughly follow the shape of the optical absorption with a peak near 0.6 eV, but the current data is insufficient to verify the existence of such a peak.

It is not possible to compare quantitatively the experimental value of $\chi^{(3)}$ with theory without taking into account the low bulk density of polyacetylene (~ 0.4) and the disordered arrangement of the polyacetylene chains. If these corrections are made, the calculated and experimental values for $\chi^{(3)}$ in the nonresonant regime are in agreement. These values are comparable to or greater than that observed for inorganic semiconductors. A slightly smaller $\chi^{(3)}$ is found for PTS. For optical switching applications, it is important that little light be absorbed in the nonlinear medium. This not only reduces the optical energy needed to operate the switch, but also limits problems caused by excessive heating. It is important to note that the intrinsic absorption coefficients for both PTS [9] and polyacetylene [10] are low in this region of the spectrum, and thus the figures of merit $\frac{\chi^{(3)}}{\alpha}$ are high.

3. PHYSICAL PROPERTIES/PROCESSABILITY.

In the family of conjugated polymers, the cumulenes and polyynes are the least studied since they share the properties of limited stability and intractability. Polyacetylenes have limited tractability, and in thin film form are unstable towards oxidation under ambient conditions. In the absence of encapsulation, their use as materials for nonlinear optics currently appears to be limited to fundamental studies. In contrast, polydiacetylenes have proven to be far more robust,

and their availability as single crystals, thin crystalline films, Langmuir-Blodgett multilayers, and as cast or spun films makes them particularly attractive materials for nonlinear optics.

Though they are versatile, polydiacetylenes are not without flaws. Polydiacetylenes are synthesized by topotactic polymerization, a process that transforms a single crystal of

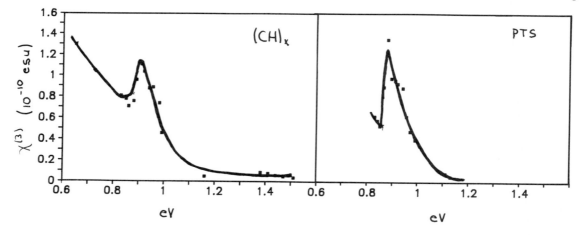

Figure 3. Spectrum of $\chi^{(3)}$ obtained from frequency tripling experiments. (a) *trans*-polyacetylene. (b) PTS.

monomer to a polymer single crystal. An increase in density accompanies this transformation, and strain induced by the polymerization can lead to defects and flaws in the crystals. X-ray topography has been used to image these defects [11] that include growth dislocations, slip dislocations, twinning, and growth sector boundaries. Direct lattice imaging of polydiacetylenes by electron microscopy [12] has also revealed chain end dislocations that form during polymerization. Obviously the task of growing defect free single crystals from solution or from the melt and the conversion of these materials to high quality polymer single crystals can be problematic.

One new approach that yields high quality, large area thin films of polydiacetylenes is the solution or melt-shear technique. [7] In this method (Figure 4), crystallization from the melt or from solution takes place between two optically smooth plates. The spacing between the plates thus determines the thickness of the resulting crystal. PTS monomer crystals have been prepared by solution evaporation, and TCDU and ETCD have been crystallized from the melt. The monomer crystals were recovered by washing them from the substrate and were polymerized either by thermal annealing or by exposure to gamma radiation.

In our work, we have observed that PTS monomer crystals that are polymerized while still clamped between the plates often fracture with a characteristic pattern (Figure 4). These parallel fractures are nearly uniformly spaced, and apparently form along crystallographic axes. The cause of the fracturing is the difference in the unit cell sizes between monomer[13] [14] and polymer.[15] When PTS monomer crystals polymerize, the unit cell shrinks by about 5% along the b axis (chain direction) while the a and c axes decrease by less than 1%. With the crystals clamped between the plates, the lattice mismatch can only be accommodated by the development of fractures perpendicular to the b axis. Since at low conversion, polymer was observed only in fractured areas, it is tempting to speculate that strain release (fracturing) induces polymerization near the crack. Since the overall dimensions of the crystals do not change during the transformation, the measurements of crack widths should suggest whether such a mechanism has merit. Indeed, optical micrographs show that the fractures in the polymer crystal correspond to about 5% of the original PTS monomer crystal.

Several soluble polydiacetylenes have been synthesized that can be treated as typical polymeric materials. These materials offer tremendous processability advantages since such techniques as the casting of films from solution and spin coating can be used to generate thin uniform films. In our hands, 4-BCMU deposited from cyclopentanone gave uniform, optically clear films ranging from 0.1 to 1 µm in thickness. Similarly, films of 3-BCMU were cast from dimethylformamide solutions. These films scatter little light, and are a convenient source of samples for optical experiments. Cast films however, do not retain the high degree of order found in the polymer crystal. The linear chains in polymer crystals must be contrasted with the more disordered arrangement of chains deposited from isotropic solutions. This difference is often manifest by a slight shift

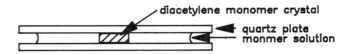

Figure 4. (top) Technique for the preparation of thin single crystals of polydiacetylene monomer. (bottom) Fractured PTS polymer crystal. The polymer chains lie perpendicular to the fractures. (250×)

in the optical absorption spectrum to higher energies, and since $\chi^{(3)}$ scales with $1/E_g$, a decrease is observed for $\chi^{(3)}$. Although both 4-BCMU and PTS are nominally infinite polydiacetylenes, the magnitude of $\chi^{(3)}$ at 1.06 μm for dried films of 4-BCMU is $\sim 1 \times 10^{-11}$ esu.[16] which is an order of magnitude less than that of PTS. [17]

4. POLYMER STABILITY

Conjugated polymers are generally susceptible to oxidation, with thin films of polyacetylene degrading in hours. Crystalline derivatives of polydiacetylenes are robust, but they are not indefinitely stable. Samples exposed to ambient room lights and oxygen photooxidize and develop discolored surfaces and a concurrent bleaching of their optic absorption spectrum. For thick crystals this degradation is barely noticeable since only a thin surface layer is affected, but optically thin crystals can badly degraded.

The photoinduced degradation of soluble polydiacetylenes has been studied.[18] Solutions of TS-12, 3-BCMU, and 4-BCMU in chloroform degraded when exposed to mid UV radiation. While only a drop in molecular weight was seen initially, eventually a loss in the strength of the optical absorption and a shift of the absorption band to higher energies followed. Both effects are deleterious to the magnitude of $\chi(3)$. Fortunately, the polydiacetylenes were stable for irradiation wavelengths > 400 nm.

5. SUMMARY

Conjugated polymers combine large third order optical nonlinearities with ultrafast response times. The topotactic polymerization of diacetylene monomers can be used to yield good quality single crystals, and soluble derivatives offer the advantages of conventional polymer processing techniques such as the spin-casting of thin, uniform films. None of the conjugated polymers are rigorously stable to oxidation, but polydiacetylenes, especially in single crystal form, are robust.

While most factors point to the diacetylenes as the material of choice in the family of conjugated polymers, it is sobering to note that the conjugated backbone that is the source of the nonlinearity is only a small fraction of the unit cell volume. Most of the unit cell is occupied by the substituents that give the polydiacetylenes such desirable features as solubility and the ability to form high quality single crystals. Thus on a volume basis alone, the $\chi^{(3)}$ of PTS would be expected to be only 1/6 that of a fully dense polyacetylene crystal.

6. ACKNOWLEDGEMENTS

Thanks to Brian Meagher for the optical micrograph, and to Sukant Tripathy for useful discussions about the solution-shear technique for growing monomer crystals.

7. REFERENCES

1. For reviews see: G. Khanarian, Ed. *Proc. SPIE-Int. Soc. Opt. Eng.* **682**, (1987); D.J. Williams, Ed. Nonlinear Optical Properties of Organic and Polymeric Materials, American Chemical Society: Washington, DC, 1983.

2. J.P. Hermann, J. Ducuing, *J. Appl. Phys.* **45**, 5100 (1974).

3. G.P. Agrawal, C. Cojan, C. Flyntzaniz, *Phys. Rev. B* **17**, 776, (1976).

4. D.N. Beratan, J.N. Onuchic, J.W. Perry, *J. Phys. Chem.* **91**, 2696, (1987).

5. F. Wudl, S. Bitler, *J. Am. Chem. Soc.* **108**, 4685 (1986).

6. F. Kajzar, S. Etemad, G.L. Baker, J. Messier *Solid State Commun.* **63**, 1113, (1987).

7. M. Thakur, S. Meyler *Macromolecules* **18**, 2341 (1985).

8. F. Kajzar, J. Messier *Thin Solid Films* **132**, 11 (1985).

9. M. Lequime, J.P Hermann *Chem. Phys.* **26**, 431, (1977).

10. B.R. Weinberger, C.B. Roxlo, S. Etemad, G.L. Baker, J. Orenstein *Phys. Rev. Lett.* **53**, 86 (1984).

11. M. Dudley, J.N. Sherwood, D.J. Ando, D. Bloor in *Polydiacetylenes*; D. Bloor, and R.R. Chance Ed.; Martin Nijhoff Publishers: Boston, 1985; NATO ASI Ser. E, No. 102.

12. R.T. Read, R.J. Young *J. Mater. Sci.,* **19**, 327 (1984).

13. D. Bloor, L. Koski, G.C. Stevens, F.H. Preston, D.J. Ando *J. Mater. Sci.* **10**, 1678, (1975).

14. P. Robin, J.P. Pouget, R. Comes, A. Moradpour *J. Phys. (Paris)* **41**, 415, (1980).

15. D. Kobelt, E.F. Paulus *Acta Crystallog. Sect. B* **30**, 232 (1974).

16. S. Etemad, G.L. Baker, unpublished results.

17. C. Sauteret, J.P. Hermann, R. Frey, F. Pradere, J. Ducuing, R.H. Baughman, R.R Chance *Phys. Rev. Lett.* **36**, 956 (1976).

18. M.A. Muller, G. Wegner *Makromol. Chem.* **185**, 1727, (1984).

Invited Paper

Modelling, synthesis, and characterization of organic
nonlinear optical materials

W.C. Egbert

Corporate Research Laboratories, 3M Company,
3M Center 201-2S-05, St. Paul, Minnesota 55144-1000

ABSTRACT

The introduction of advanced optical materials into communication technology facilitates increased data transmission speed, increased bandwidth, intrinsic low crosstalk and high security in multiplexed systems, and simplified implementation of parallel processing. Prototype and first-generation optical communication, information storage and information processing systems have been developed from hybrid electro-optical or all-optical effects in inorganic crystals and semiconductors. Organic and polymer nonlinear optical materials may have processing or performance advantages in the next generation of optical information systems design, but research is needed to define the materials of interest and development will be necessary to incorporate these new materials into practical active elements. This paper is an overview of 3M's program for development of nonlinear optical materials.

1. INTRODUCTION

The potential for high performance organic and polymer nonlinear optical (NLO) materials makes them attractive for the next generation of NLO applications. At this time, organic materials are used in optical devices primarily as passive agents, e.g., adhesives and index-matching agents. Large NLO susceptibilities have been reported for organic crystals and thin films, and there is reason to believe that these large nonlinearities may provide the basis for new optical devices. The technical challenge will be to maintain optical performance in useful device configurations which are straightforward to process and implement.

We are developing organic and polymer materials with optimized optical nonlinearities for incorporation into optical devices for communications, information storage and information processing systems. 3M's program includes computer modelling of chemical structures and their relation to optical properties, synthesis of new compounds directed in part by the computer modelling activity, characterization of the optical nonlinearity of these new compounds, and attention to the practical aspects of implementing these new materials in functional prototype device demonstrations. An example of a nonlinear optical material developed at 3M will illustrate this unified approach to materials development.

2. MOLECULAR MODELLING

The first step is selection of molecular and polymeric systems with potentially large optical susceptibilities from the many possible classes of organic compounds. This selection process begins with the the physical theory of NLO processes.[1] The frequency-dependent polarization response of a medium to an applied electric field may be written as a function of the static polarization P^o, linear susceptibility $\chi^{(1)}$, and nonlinear susceptibilities $\chi^{(2)}$ and $\chi^{(3)}$:

$$
\begin{aligned}
P_i(\omega) = & P_i^o + \chi_{ij}^{(1)}(-\omega;\omega)E_j(\omega) \\
& + \chi_{ijk}^{(2)}(-\omega;\omega_1,\omega_2)E_j(\omega_1)E_k(\omega_2) \\
& + \chi_{ijkl}^{(3)}(-\omega;\omega_1,\omega_2,\omega_3)E_j(\omega_1)E_k(\omega_2)E_l(\omega_3) + \cdots
\end{aligned}
$$

$$(1)$$

This expression for the polarization implicitly includes useful boundary conditions for NLO materials. (The usual convention for summation over indices is assumed). First, the polarization is a macroscopic observable, by its radiation to the output optical field. Second, the presence of the linear term $\chi^{(1)}$ implies that low polarizability and consequently low dielectric constant at the frequency of interest will minimize the effect of internal depolarizing (local) fields. Third, the frequency ranges of interest may run from dc to optical frequencies, as determined by the application, and low dielectric constants (small $\chi^{(1)}$) will reduce the response time. Finally, the hyperpolarizability tensors are governed by the symmetry of the medium, so the second rank tensor $\chi^{(2)}$ is identically zero in materials with a center of inversion symmetry. These points define the boundary conditions for materials development, but they are insufficient to guide the selection and synthesis of NLO organic and polymeric materials.

Organic and polymeric materials are typically ensembles of small units, either molecules in a crystal or active moities attached as pendant groups to a polymer backbone. (Macromolecular systems and polymers with highly delocalized nonbonding π-electrons, e.g., one dimensional organic conductors and polydiacetylene derivatives, present special theoretical and experimental challenges and will be considered in a future paper). The macroscopic hyperpolarizability of a molecular crystal or polymer matrix is an ensemble average of the microscopic molecular susceptibility β, taking into account the number N of active molecules, the rotation matrix $\boldsymbol{R}(\theta,\phi,\psi)$ which transforms the laboratory frame into the molecular principal axes, and the Lorentz local field correction factors f. For the second order susceptibility $\chi^{(2)}$, the expression becomes [2]:

$$
\chi_{ijk}^{(2)} = N\langle \boldsymbol{R}_{im}\boldsymbol{R}_{jn}\boldsymbol{R}_{kp}f_{mq}^{\omega}\beta_{qrs}(-\omega;\omega_1,\omega_2)f_{nr}^{\omega_1}f_{ps}^{\omega_2}\rangle \quad (2)
$$

A similar expression relates the third order hyperpolarizability $\chi^{(3)}$ to the third order molecular susceptibility γ. As noted above, the invariance of $\chi^{(2)}$ under the symmetry operations of the medium restricts nonzero $\chi^{(2)}$ to noncentrosymmetric systems; no symmetry restrictions apply to $\chi^{(3)}$.

In principle one could measure the molecular hyperpolarizabilities β and γ directly, apply the local field corrections (if known) and coordinate rotations, and calculate the macroscopic susceptibility. One might infer from Equation 2 for $\chi^{(2)}$ and its counterpart for $\chi^{(3)}$ that incorporating materials into appropriately oriented structures reduces to a three dimensional geometry problem. However, the local field correction factors f in Equation 2 are phenomenological parameters, not easily measured in situ, especially in ultrafast (femtosecond) time regimes which are appropriate for purely electronic excitations of the nonbonding electronic wavefunctions in an organic NLO material. Measurements made in solution may not carry over to the solid state, because intermolecular interactions in the solid state will modulate local fields in ways which are not amenable to numerical calculation. Therefore, it is generally more practical to design molecules for large nonlinear susceptibility as described below, model the interactions in the macromolecular structure, then measure the macroscopic response of the crystal or polymer matrix which is organized to give the maximum alignment of the major optical axis of the active moiety. As the structure-property relationships for molecules and molecular organizates are better understood, it will be feasible to design a NLO material, starting at the molecular level and using first principles at each subsequent step.

The expression for the macroscopic polarization, Equation 1, by itself offers little guidance for the synthesis of new molecular NLO materials. As a consequence, several general rules have been developed and stated in terms which give some direction to synthetic chemistry efforts. First, extended systems of conjugated bonds have been demonstrated to have large NLO susceptibility in molecules and polymers.[3] Second, charge transfer from donor to acceptor substituents within a molecule generally enhances the conjugation. This is an imperfect criterion - too strong a transfer tends to localize the charge and reduce the molecular susceptibility, with the associated disadvantage that highly polarized molecules tend to dimerize.[4-6] This is illustrated by 4-nitroaniline, a prototypical nonlinear organic molecule. Calculations on 4-nitroaniline indicate that the nonlinear response of the isolated molecule approximates that of a two level system.[7] The formalism of perturbation theory applied to the two level system gives an expression which suggests a third molecular design criterion, viz., $\chi^{(2)}$ is enhanced by maximizing the difference between the ground and excited state dipoles and by maximizing the transition dipole matrix element from the ground to first excited state.[1] These three rules for increasing molecular nonlinear susceptibility - charge conjugation, charge transfer, and maximizing transition dipoles - do not take into account the effects of intermolecular interactions or applied electric fields. These simple rules do not predict with certainty the effect of substituent changes, crystallization, and polymerization on the optical nonlinearity, since any one of

these processes may alter the π-electron distribution in the conjugated bond network of the constituent molecules. A theoretical framework which models the π-electron distribution responsible for optical nonlinearity of molecular organizates and predicts the effect of chemical substitutions would be more useful than a simple molecular model.

We have developed a technique for the visualization of the origin of optical nonlinearity in organic molecules which does not depend on explicit calculation of the ground state, excited state, or transition dipole matrix elements. Quantum chemical calculations of these matrix elements in extended molecules (large dyes or polymer repeat units, for example) become unmanageable because of CPU time constraints and disk storage limits for intermediate results. These limitations on a direct "brute force" calculation encouraged us to develop an alternative approximation which reasonably models the effect of molecular substitution on the optical nonlinearity. In our approach, the nonbonding molecular orbitals are calculated by extended Hückel or other approximations.[7] The essential criterion is that the intermediate virtual state of the nonbonding π-electrons responsible for the susceptibility is coupled through the dipole matrix elements to the initial and final states, and this coupling occurs only when there is a shared element of symmetry among the initial, intermediate, and final state. The results give us a qualitative estimate of the magnitude of the optical nonlinearity, the axes of highest molecular hyperpolarizability, a visual representation of the effect of substitutions on the core molecule, and the effect of intermolecular interactions in the solid state.

The degree of overlap between the ground and virtual excited state(s), and the extent of π-electron delocalization, determines the relative size of the molecular susceptibility. Generally, though not always, the highest occupied molecular orbital (HOMO) and lowest unoccupied molecular orbital (LUMO) are the only orbitals needed, i.e., the two-level model is sufficient. For those cases in which the two level system is not adequate, higher lying orbitals may be included, with the proviso that they share an element of symmetry with the HOMO. The required molecular orbital overlap may often be determined by inspection. This formulation allows interpretation of the effect of the chemistry on the NLO response in computer-generated molecular structures. These structures may be available from direct measurement, e.g., x-ray crystallography, or they may be theoretical proposals for new materials. For those cases which are not obvious by inspection, the model provides a rationale for breaking down the required calculations into manageable units, linked to provide the final answer.

We used extended Hückel molecular orbital calculations to determine the stationary state electron wavefunctions.[8] We have found that the eigenvalues calculated by this method need not be exact to represent the symmetry of the HOMO and LUMO. The effect of a polarizing electric field is approximated by taking the sum and difference of the HOMO and LUMO to produce an intermediate mixed state approximating the virtual excitation of the molecule by the photon fields of the nonlinear process. The symmetry-controlled coupling between molecular states is similar to the Woodward-Hoffman rule for organic chemical reaction theory, i.e., the intermediate state in a reaction shares an element of symmetry with the initial and final states, and this shared element of symmetry defines the reaction pathway in phase space. Most importantly from the perspective of the synthetic chemist, this modelling technique gives a graphic depiction of the effect of chemical substitution on the π-electron delocalization. The π-electron delocalization affects the polarizabilities of the relevant states which are responsible for the molecular susceptibility and defines the axes for coupling to intermediate states in the NLO process. Chemical substitutions which affect the π-electron distribution therefore control the magnitude and orientation of the axes of molecular susceptibility, and the molecular orbital model represents these effects.

These points are illustrated by the mixing of HOMO and LUMO for two nitroaniline derivatives. Figures 1 and 2 show the HOMO, LUMO, and mixed intermediate states of 4-nitroaniline and 3-nitroaniline, respectively. The π-electron densities are indicated by the relative size of the "clouds" at each atomic site. The combination wavefunctions on the right of Figures 1 and 2 are generated by a nonresonant optical field, with energy equivalent to the midpoint of the bandgap between HOMO and LUMO, polarized with its electric vector along the major axes of the HOMO and LUMO for 4-nitroaniline, and along the bisector between the amino- and nitro- groups (1- and 3- positions) in 3-nitroaniline. These examples illustrate the utility of the molecular orbital graphical representation. The orbital structure of 4-nitroaniline is easily understood as a fully conjugated delocalized π-electron distribution, as is reflected in the combination wavefunction. The degree of asymmetry in the π-electron distribution of the mixed state as the electric field reverses direction is a measure of the anharmonic potential which gives rise to harmonic generation. The donor and acceptor in 3-nitroaniline are not conjugated in the usual sense of "starring" positions to describe the resonant structures. Nonetheless, the combination wavefunctions in Figure 2 clearly show the asymmetric polarization of the π-electron distribution as the optical electric field reverves direction each half cycle. Keeping in mind that the extended Hückel calculation is approximate, the degree of asymmetry evident in the combination wavefunctions in Figures 1 and 2 is consistent with the roughly 3:1 ratio of molecular susceptibilities β measured in electric field induced second harmonic generation (EFISH) experiments on 4-nitroaniline and 3-nitroaniline, respectively.[3,9,10] This illustrates the use of the symmetry-based model to determine the effect of substitution at a particular site in the molecule and to provide a ranking of NLO activity within classes of molecules.

As an example of the role of molecular modelling, consider the class of 1,3-substituted-2,4,6-(1H,3H,5H)-pyrimidinetriones, also known as N,N'-substituted barbituric acids. The basic structure of these compounds is analogous to that of urea; see Figure 3. The HOMO and LUMO π-electron distributions, Figure 4, of N,N'-dimethyl barbituric acid (DMBA) share a plane of symmetry along the plane bisecting the molecule. The electron distribution shifts from the acceptor-like sites in the HOMO to the donor-like sites in the LUMO.

The crystal structure of DMBA, Figure 5, shows evidence of hydrogen bonding along the direction of the molecular axis. This head-to-tail orientation maintains the noncentrosymmetric structure which is necessary for second order optical nonlinearity. The planar structure is repeated in macroscopic crystals of DMBA, which grow to large (cm^3) rhombohedral plates.

3. SYNTHESIS

Molecular modelling is a useful way to identify the compounds which need to be synthesized to find those with suitably large optical nonlinearity. One drawback to highly polarized electron donor-acceptor molecules such as 4-nitroaniline, Figure 1, is that the strong ground state dipole leads to dimer formation. Thus, in the case of 4-nitroaniline, the addition of a methyl group to form 2-methyl-4-nitroaniline does not greatly modify the susceptibility of the basic 4-nitroaniline molecule, but steric hindrance forces a noncentrosymmetric crystal structure and nonvanishing second order macroscopic hyperpolarizability $\chi^{(2)}$.[11] As can be seen in Figure 1, the 2-position in the 4-nitroaniline π-electron distribution is a region of low π-electron density, even in the polarized intermediate (mixed) state. The additional methyl group will therefore be weakly coupled to the nonlinear optically active π-electrons and consequently have little effect on the NLO properties, as is observed. The additional methyl group also inhibits dimer formation.

The HOMO and LUMO electron density distributions for DMBA, Figure 4, suggests that substitutions at the N,N' positions will have a small effect on the orbitals which are responsible for the molecular susceptibility. To test this hypothesis, we examined the effect of several substitutions, either by condensation of the appropriately di-substituted urea with malonic acid, or by alkylation of DMBA. The series of substitutions are listed in Table 1, which also lists the second harmonic generation (SHG) efficiency of the powdered samples, referenced to a urea standard.[12,13] The powder SHG test will be discussed in the next Section.

The substituted barbituric acids listed in Table 1 highlight an important aspect of organic NLO materials development. The core molecular structure responsible for the optical functionality is separable from the factors which control the physical structure of the ensemble, whether it be a crystal or macromolecule. The crystal structure of a substituted barbituric acid is determined in part by hydrogen bonding in the solid state. In Figure 5, the head-to-tail hydrogen bonding in DMBA is sketched in. The parent N,N' barbituric acid (with hydrogens attached to the nitrogens) dimerizes into a centrosymmetric pair of molecules, Figure 6. These dimers have an obvious center of inversion symmetry, and the N,N' barbituric acid has identically zero

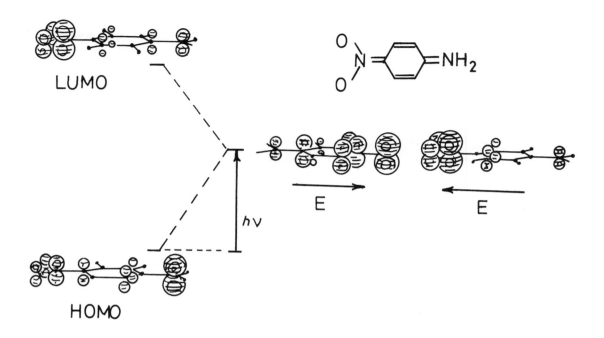

Figure 1. 4-nitroaniline HOMO, LUMO, and mixed state.

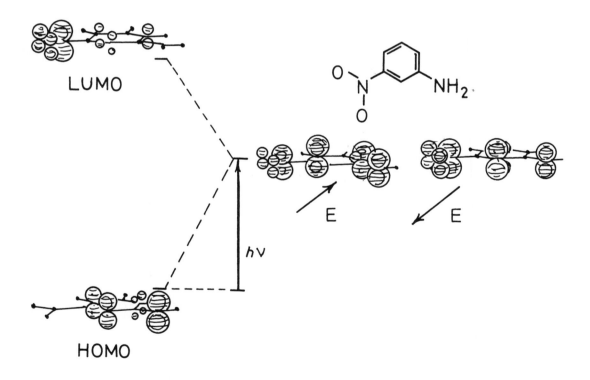

Figure 2. 3-nitroaniline HOMO, LUMO, and mixed state.

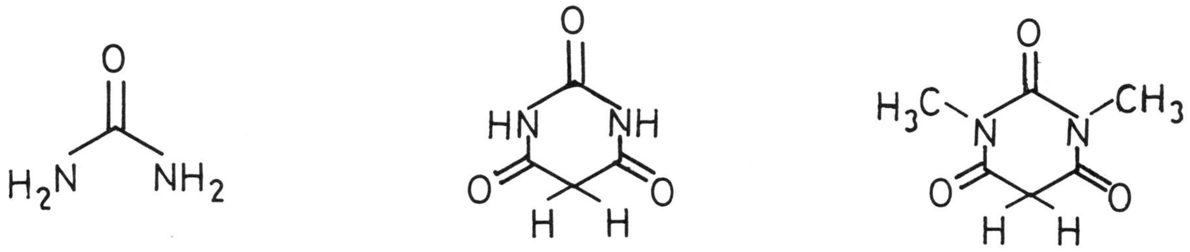

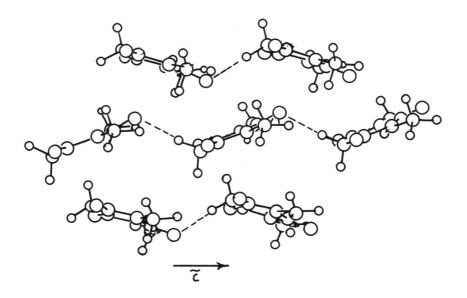

UREA

BARBITURIC
ACID

DIMETHYL
BARBITURIC
ACID

Figure 3. Urea and barbituric acids.

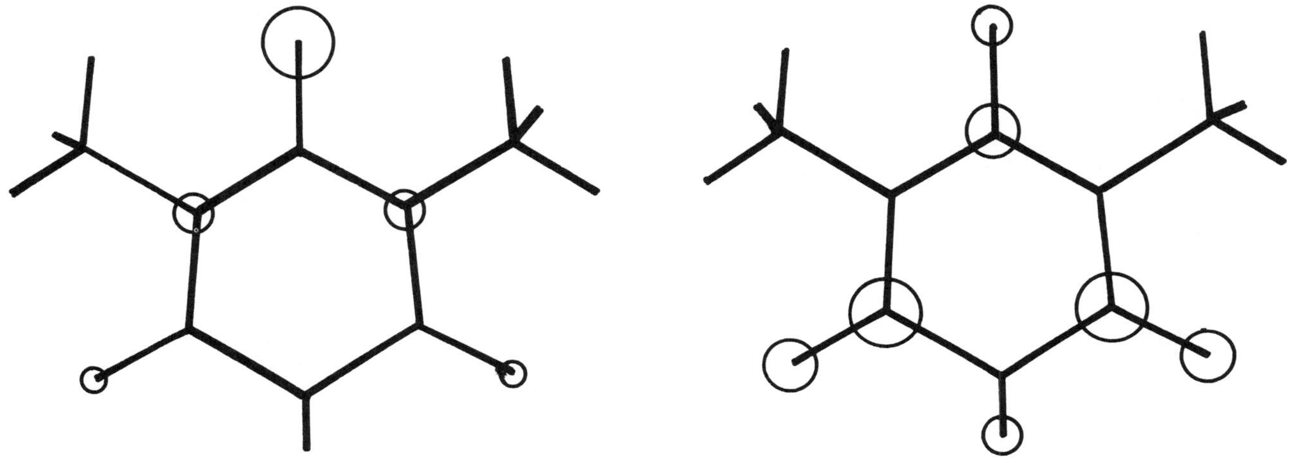

HOMO

LUMO

Figure 4. 1,3-dimethyl barbituric acid HOMO and LUMO.

$\tilde{c}$

Figure 5. 1,3-dimethyl barbituric acid crystal structure.

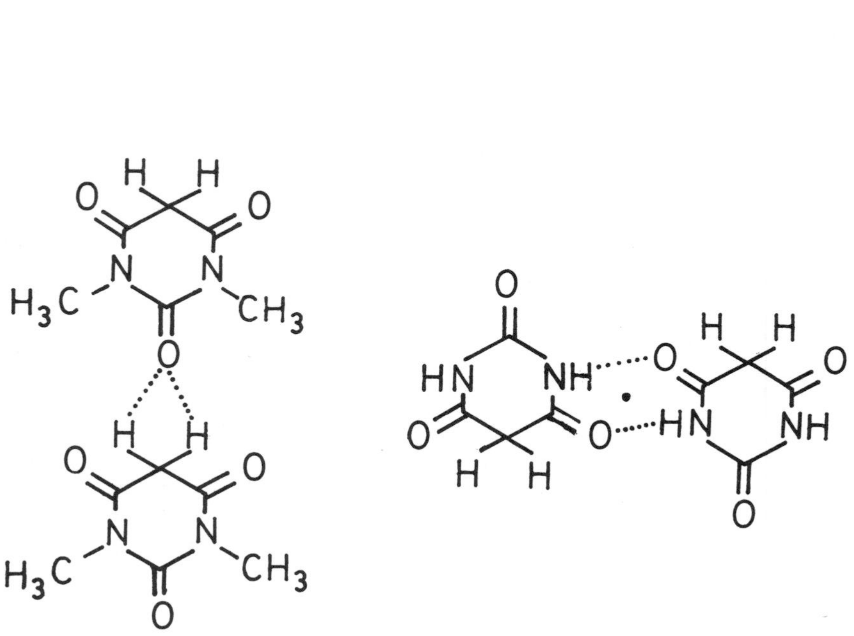

Figure 6. Hydrogen bonding in barbituric acid
and dimethyl barbituric acid.

Table 1.   Powder test second harmonic generation efficiency
of some N,N'-substituted barbituric acids.

	R_1	R_2	R_3	R_4	Relative efficiency η (urea $\eta = 1$)
(dimethyl)	CH_3	CH_3	H	H	3
(diethyl)	C_2H_5	C_2H_5	H	H	3
(n-propyl)	$n\text{-}C_3H_7$	$n\text{-}C_3H_7$	H	H	< 0.001
(i-propyl)	$i\text{-}C_3H_7$	$i\text{-}C_3H_7$	H	H	0.004
(nitrophenyl)	$C_6H_4NO_2$	$C_6H_4NO_2$	H	H	1.3
	CH_3	C_6H_5	H	H	< 0.001
	CH_3	CH_3	CH_3	H	< 0.001
	CH_3	CH_3	C_7H_7	C_7H_7	0.002

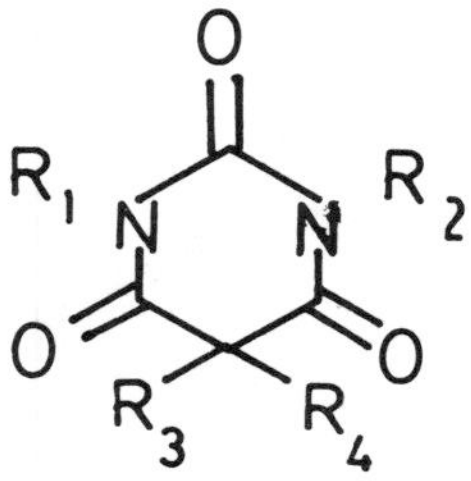

macroscopic hyperpolarizability $\chi^{(2)}$. In contrast, the 1,3-dimethyl barbituric acid, as noted above, hydrogen bonds through the oxygen at the 2 and hydrogens at the 5 positions in the solid state since the 1,3 positions are blocked by the methyl groups. This results in the head-to-tail bonding, as shown in Figure 6. The lack of side-to-side hydrogen bonding and steric (space-filling) factors results in a non-centrosymmetric structure for DMBA.

Note in Table 1 the high SHG powder test efficiency (compared to the urea standard) for the methyl- and ethyl-substituted barbituric acids and correspondingly low efficiency of all but the nitrophenyl-substituted barbituric acids. The data and modelling available on these compounds indicate that the low efficiencies have their source in nonoptimal crystal packing which is to varying degrees centrosymmetric, rather than intrinsically low molecular susceptibility. This illustrates the use of molecular modelling to provide lists of candidate materials and their expected properties, even before the chemical synthesis is undertaken. It also emphasizes the use of modelling to understand the source of low hyperpolarizabilities, whether it be the basic structure of the molecule giving small molecular susceptibility or nonoptimum macroscopic structure of the lattice.

4. OPTICAL CHARACTERIZATION

The computer modelling described in Section 2 above is used to narrow the choices of materials for synthesis. Together with several simple screening tests, large numbers of new materials can be examined for NLO properties with a minimum of sample preparation and post-test data analysis. This combined modelling and testing quickly identifies the molecular systems which have reasonably large nonlinear susceptibility, as well as their mechanism of operation. It also identifies those systems with large hyperpolarizability suitable for in-depth characterization and development.

Depending on the material form, one of several screening tests may be used to test new materials. Crystalline powders are tested by powder SHG, a simple and useful way to rapidly screen large numbers of materials for second order hyperpolarizability. By varying the wavelength of the input laser and the particle size in the sample, the powder SHG screening test is used to determine the spectral regions of phase matching, i.e., phase coherence between the fundamental and second harmonic, which is a function of the dispersion in the optical indicatrices at the fundamental and harmonic frequencies.[12,13] The efficiencies for powder SHG of barbituric acid derivatives reported in Table 1 are compared to a urea standard.[14] Dimethyl barbituric acid is phase matchable for 1064 nm, but not for 750 nm, fundamental. The probable cause of the loss of phase matching is a π-π^* transition at 240 nm, with a tail extending to 350 nm, which would affect the index of refraction for the second harmonic of 750 nm. The models can also be used to select materials for screening for third order nonlinearity, using intensity-dependent index of refraction (self-focusing or defocusing) compared to a carbon disulfide standard.[15]

For those cases where modelling indicates a molecular second order susceptibility β, but powder SHG gives a null result, the problem may be centrosymmetric packing in the naturally-occurring crystal structure. If this is the case, artificial acentric order may be introduced by the application of an electric field to the material in the molten state or in solution. In electric field induced second harmonic generation (EFISH), the product of the intrinsic dipole μ and the molecular susceptibility tensor β is measured.[16] A limitation of EFISH is illustrated by the barbituric acids listed in Table 1. Their ground state dipole moments, as determined by modelling, are small and will not maintain molecular orientation in solution. An alternative method of applying artifical noncentrosymmetric order is the use of Langmuir-Blodgett techniques, where the molecule is modified to a surfactant by the addition of a nonpolar tail on the polar headgroup.[17] This is a subject of current study in our program and will be reported in depth at a later date.

The screening tests establish which materials are suitable for further development, including accurate determination of the hyperpolarizability ($\chi^{(2)}$ or $\chi^{(3)}$) tensor elements. Powder SHG (Table 1) shows that microcrystalline dimethyl barbituric acid has a large $\chi^{(2)}$. DMBA grows into reasonably large single crystals from solution, so it is a suitable candidate for detailed determination of $\chi^{(2)}$ by the Maker fringe technique.[18] In this technique the effective optical path length through an oriented single crystal sample is varied, either by rotating a plane parallel slab about an axis normal to the test laser beam, or by translating a wedged sample across the beam. As the path length varies, the wavevector mismatch between the fundamental and the second harmonic caused by operating off of the phase-matching direction leads to fringes as the optical path length sweeps through the envelope function $\sin^2(\Delta k \cdot L/2)$. The interference fringes in the second harmonic signal can be compared to the fringes generated by a calibration sample, e.g., alpha-quartz, to determine the magnitude of individual tensor components. The result for

DMBA is $\chi^{(2)}_{zzz} = 1.58 \times 10^{-9}$ esu, about the same as d_{11} for alpha-quartz. From the molecular two level model, described above, one would expect a $\chi^{(2)}_{zzz}$ from the mixing of the HOMO and LUMO shown in Figure 4. The two level model does not predict the existence of a large $\chi^{(2)}_{zxx}$, but by inspection of the molecular structure it is apparent that this would represent excitation across the 2,5 position (the N,N'-substitutions). The simple two-level model is sufficient for a first approximation, but the higher-lying unoccupied molecular orbital states may be significant to the understanding of the other optical excitations of DMBA.

5. SUMMARY

3M's program incorporates molecular modelling, chemical synthesis, and nonlinear optical characterization. The molecular modelling is based on the representation of symmetry-controlled interactions among nonbonding π-electron molecular orbitals. The modelling is used to predict the effects of chemical derivatization and to select the most promising substitutions for work-up. The optical characterization consists of rapid screening tests to examine many materials to quickly select out the best materials suitable for in-depth study and incorporation into prototype devices or other demonstrations of feasibility.

In this introduction to 3M's program we have used the family of barbituric acid derivatives as a representative example in which molecular modelling, synthetic chemistry, and optical characterization have resulted in a new class of materials for nonlinear optical effects. We have not discussed the delocalization of electrons in long-chain materials, e.g., diacetylene polymers or organic conductors, nor have we discussed third order nonlinearities, except in passing; our example system, the barbituric acids, were not developed as $\chi^{(3)}$ materials and have not been evaluated for $\chi^{(3)}$ effects.

I would like to acknowledge the assistance of several members of the Advanced Optical Materials group in the preparation of this overview. Dr. David Ender provided the single crystal measurements on DMBA and Carl Nelson performed the powder SHG tests. Ellen Harelstad synthesized the family of barbituric acids. Dr. Peter Leung determined the crystal structure and generated the HOMO-LUMO electron distribution maps of DMBA. The theoretical approach was developed by Dr. John Stevens who also provided interpretation of modelling versus test results.

6. REFERENCES

1. Y.R. Shen, **The Principles of Nonlinear Optics** (Wiley, New York, 1984).

2. A.F. Garito, K.Y. Wong, Y.M. Cai, H.T. Man, and O. Zamani-Khamiri, "Fundamental nonlinear optics issues in organic and polymer systems," Molecular and Polymeric Optoelectronic Materials: Fundamentals and Applications, Garo Khanarian, Editor, Proc. SPIE **682**, 2-11 (1986).

3. B.F. Levine and C.G. Bethea, "Second and third order hyperpolarizabilities of organic molecules," J. Chem. Phys. **63**, 2666-2682 (1975).

4. J.L. Oudar and H. LePerson, "Second-order polarizabilities of some aromatic molecules," Opt. Commun. **15**, 258-262 (1975).

5. K. Jain, J.I. Crowley, G.H. Hewig, Y.Y. Cheng, and R.J. Twieg, "Optically non-linear organic materials," Optics and Laser Tech. **13**, 297-301 (1981).

6. D.J. Williams, "Organic polymeric and non-polymeric materials with large optical nonlinearities," Angew. Chem. Int. Ed. Engl. **23**, 690-703 (1984).

7. J.A. Morrell and A.C. Albrecht, "Second-order hyperpolarizability of p-nitroaniline calculated from perturbation theory based expression using CNDO/S generated electron states," Chem. Phys. Lett. **64**, 46-50 (1979).

8. J. Stevens and D.A. Ender, "Orbital symmetry considerations in the design of molecular electro-optical materials," Advances in Materials for Active Optics, Proc. SPIE **567**, 66-71 (1985).

9. B.F. Levine and C.G. Bethea, "Effects on hyperpolarizabilities of molecular interactions in associating liquid mixtures," J. Chem. Phys. **65**, 2429-2438 (1976).

10. C. Flytzanis, "Trends in optical nonlinearities of conjugated molecules and polymers," **Nonlinear Behaviour of Molecules, Atoms and Ions in Electric, Magnetic or Electromagnetic Fields**, 185-207 (Elsevier, Amsterdam, 1979).

11. B. F. Levine, C. G. Bethea, C. G. Thurmond, R. T. Lynch, and J.L. Bernstein, "An organic crystal with an exceptionally large optical second harmonic coefficient: 2-methyl-4-nitroaniline," J. Appl. Phys. **50**, 2523-2527 (1979).

12. S.K. Kurtz and T.T. Perry, "A powder technique for the evaluation of nonlinear optical materials," J. Appl. Phys. **39**, 3798-3813 (1968).

13. A. Graja, "Second harmonic of light generation in crystal powders," Acta Phys. Polonica **A37**, 539-558 (1970).

14. J.M. Halbout, S. Blit, W. Donaldson, and C.L. Tang, "Efficient phase-matched second-harmonic generation and sum-frequency mixing in urea," IEEE J. Quant. Electron. **QE-15**, 1176-1180 (1979).

15. R.C.C. Leite, R.S. Moore, and J.R. Whinnery, "Low absorption measurements by means of thermal lens effect using an He-Ne laser," Appl. Phys. Lett. **5**, 141-143 (1967). See also reference 1, Chapter 17.

16. C.C. Teng and A.F. Garito, "Dispersion of the nonlinear second-order optical susceptibility of organic systems," Phys. Rev. **B28**, 6766-6773 (1983).

17. For a recent review see A. Barraud and M. Vandevyver, "Growth and characterization of organic thin films (Langmuir-Blodgett films)," Ch. II-5 in **Nonlinear Optical Properties of Organic Molecules and Crystals Vol. I**, D.S. Chemla and J. Zyss, eds. (Academic Press, Orlando, FL, 1987).

18. F. Zernike and J. Midwinter, **Applied Nonlinear Optics**, 82-83 (Wiley, New York, 1973).

Optical second harmonic generation from Langmuir-type molecular monolayers

G. Berkovic,[*] Th. Rasing, and Y. R. Shen

Department of Physics, University of California
Center for Advanced Materials, Lawrence Berkeley Laboratory
Berkeley, California 94720

ABSTRACT

A single molecular layer is generally sufficient to produce observable optical second harmonic generation (SHG). Furthermore, the selection rules governing this process make the SHG from a single monolayer often stronger than that from the medium supporting the monolayer. We have studied SHG from various Langmuir-type monolayers (i.e. monolayers spread on a water surface) in the following contexts:
(1) Study of chemical reactions (e.g. polymerization) and two-dimensional phase transitions in molecular monolayers on water.
(2) Development of a new technique to evaluate optical nonlinear coefficients of organic molecules, and their relationship to the molecular structure.

1. INTRODUCTION

Insoluble molecular monolayers at liquid-gas or liquid-liquid interfaces have been the subject of numerous studies in many fields of basic and applied research. They are rather ideal for studying two-dimensional phase transitions[1] and as model systems for biological membranes. Polymerizable monolayers have been employed as ultrathin coatings in microlithography[2] and microelectronics.[3] Potential applications of organic molecules with very large optical nonlinear coefficients for optoelectronic devices[4] have set off intense efforts in finding such materials and to incorporate them in Langmuir-Blodgett type structures or thin polymeric films.

In all these cases, characterization of the structure and/or optical nonlinearity of the molecular monolayer is essential in order to understand its properties and to be able to custom design new materials and structures. For this purpose analytical tools are needed that are able to detect, nondestructively, a monolayer or less at the surface of a substrate. Though there has been some recent progress in infrared[5] and x-ray diffraction[6] analysis there is a clear need for additional experimental techniques for the study of molecular adsorbates at interfaces.

In this paper we will demonstrate how we can use optical second harmonic generation (SHG) to study the structure, optical nonlinearities and polymerization of Langmuir films of organic molecules. All the materials studied are spreadable on a water surface, which has the advantage of a low background SHG level and an easily controllable surface density of adsorbate molecules.

The effectiveness of SHG as a surface probe stems from the fact that in the electric dipole approximation SHG is forbidden in centrosymmetric media but necessarily allowed at a symmetry breaking interface or surface.[7] As an optical probe it has the advantage of a high spectral and time resolution and of being applicable to any interface accessible by light.

Second harmonic generation arises from the induced nonlinear polarization $\vec{P}(2\omega)$ given by

$$\vec{P}(2\omega) = \overset{\leftrightarrow}{\chi}{}^{(2)}\vec{E}(\omega)\vec{E}(\omega), \tag{1}$$

where $\overset{\leftrightarrow}{\chi}{}^{(2)}$ is a second order nonlinear susceptibility and $\vec{E}(\omega)$ the incident laser field. The surface nonlinear susceptibility $\overset{\leftrightarrow}{\chi}_s$ which is responsible for the SHG at an interface generally reflects the properties of the surface layer and can be written as

$$\overset{\leftrightarrow}{\chi}_s^{(2)} = \overset{\leftrightarrow}{\chi}_w^{(2)} + \overset{\leftrightarrow}{\chi}_m^{(2)} + \overset{\leftrightarrow}{\chi}_{int}^{(2)} , \tag{2}$$

where $\overset{\leftrightarrow}{\chi}_w^{(2)}$ and $\overset{\leftrightarrow}{\chi}_m^{(2)}$ are the susceptibilities of the substrate (here water) and adsorbate monolayer, respectively, and $\overset{\leftrightarrow}{\chi}_{int}^{(2)}$ includes any perturbational interaction between them. Although zero under the electric dipole approximation, a weak quadrupole effect causes $\overset{\leftrightarrow}{\chi}_w^{(2)}$ to be non-zero. For adsorbates of moderate nonlinearity the SHG signal from an adsorbate covered surface far exceeds that of the bare surface, and in those cases we may approximate $\overset{\leftrightarrow}{\chi}_s^{(2)} = \overset{\leftrightarrow}{\chi}_m^{(2)}$. When $\overset{\leftrightarrow}{\chi}_m^{(2)}$ is small, the substrate background cannot be neglected and in order to subtract it, one has to know the phase difference between $\overset{\leftrightarrow}{\chi}_s^{(2)}$ and $\overset{\leftrightarrow}{\chi}_m^{(2)}$. This was obtained by interference of both the bare and covered substrate signals with that from a quartz plate excited by the same pump beam. In the present case we have neglected any interaction, i.e. $\overleftrightarrow{\chi}_{int}^{(2)} = 0$.

For a surface density of adsorbates N_S and neglecting local field corrections, $\overleftrightarrow{\chi}_m^{(2)}$ can be written as

$$\chi_{m,ijk}^{(2)} = N_S \langle T_{ijk}^{\lambda\mu\nu} \rangle \alpha_{\lambda\mu\nu}^{(2)} , \qquad (3)$$

where $T_{ijk}^{\lambda\mu\nu}$ describes the coordinate transformation between the molecular (ζ, n, ξ) system and the lab (x,y,z) system, $\overleftrightarrow{\alpha}^{(2)}$ is the molecular nonlinear polarizability and the angular brackets $\langle\ \rangle$ denote an average over all molecular orientations.

The SH intensity generated from a monolayer covered surface in the reflected direction in air is given by[8]

$$I(2\omega) = \frac{32\pi^3\omega^2}{c^3\varepsilon(\omega)\varepsilon^{1/2}(2\omega)} |\vec{e}_{2\omega} \cdot \overleftrightarrow{\chi}_S^{(2)} \vec{e}_\omega \vec{e}_\omega|^2 I^2(\omega) , \qquad (4)$$

where $\vec{e}_\Omega = \overleftrightarrow{L}_\Omega \hat{e}_\Omega$, with $\hat{e}_\Omega$ denoting the unit polarization vector of the field at frequency Ω and $\overleftrightarrow{L}_\Omega$ the Fresnel factor for the field, and $I(\omega)$ the laser intensity. From Eqs. (3) and (4) we see that from the SH measurements we can deduce $\chi_{S,ijk}^{(2)}$ and $\chi_{m,ijk}^{(2)}$ and in that way obtain information about the molecular orientation (via $\overleftrightarrow{T}$ and $\overleftrightarrow{\alpha}^{(2)}$). Thus, in case of a chemical reaction resulting in a change in $\overleftrightarrow{\alpha}^{(2)}$ we can monitor this reaction in situ. Unfortunately, $\overleftrightarrow{T}$ can be quite complicated for a general molecular structure, and quantitative results might be hard to obtain. However, when $\overleftrightarrow{\alpha}^{(2)}$ is dominated by a single component $\alpha_{\xi\xi\xi}^{(2)}$ along the molecular ξ-axis and the latter is randomly distributed around the surface normal, the situation is greatly simplified. For the nonvanishing components of $\overleftrightarrow{\chi}_m^{(2)}$ one can then write:[9]

$$\chi_{m,zzz}^{(2)} = N_S \langle \cos^3\theta \rangle \alpha_{\xi\xi\xi}^{(2)} \qquad (5)$$

$$\chi_{m,zii}^{(2)} = \chi_{m,izi}^{(2)} = \chi_{m,iiz}^{(2)} = 1/2 N_S \langle \sin^2\theta \cos\theta \rangle \alpha_{\xi\xi\xi}^{(2)} \qquad i = x,y$$

where θ is the polar angle between $\hat{\xi}$ and the surface normal $\hat{z}$. For an orientationally ordered Langmuir film the orientational distribution is expected to sharply peak in a certain direction. Approximating the distribution by a δ-function we can then find a value of θ from the ratio of $\chi_{m,zzz}^{(2)}$ and $\chi_{m,zii}^{(2)}$.[10] Consequently, from the absolute value of $\chi_{m,zzz}^{(2)}$ (measured against a standard reference) the polarizability $\alpha_{\xi\xi\xi}^{(2)}$ can be obtained.[11,12]

We have applied this technique to study the change in molecular orientation at a two-dimensional phase transition,[13] to obtain the second-order nonlinearities of a series of organic molecules[11,12] and to study the polymerization of a monolayer of monomers.[14] All the monolayers were prepared by spreading solutions of the molecules on a Langmuir trough made out of glass. A moveable barrier controls the surface density of the molecules and a platinum float (Wilhelmy plate) attached to a balance measures the surface tension. For the SHG measurements we used the frequency doubled output of a Q-switched Nd:YAG laser at 532 nm as the pump beam. The surface SH signal was calibrated against the SH signal from a thin quartz plate with a bulk nonlinearity $\chi_{xxx}^{(2)} = 2.2 \times 10^{-9}$ esu.[15] Figure 1 shows a schematic picture of the experimental apparatus. With the input and output polarizers we select different $\overleftrightarrow{\chi}_S^{(2)}$ components, whereas the color filters ensure that we only observe the SH signal from the water-air interface of interest.

2. RESULTS AND DISCUSSION

Figure 2 shows the measured surface pressure, π, as a function of the surface area per molecule, A, for a monolayer of pentadecanoic acid (PDA) on a pH = 2 water substrate at various temperatures.[13] The sharp kink in the middle of each π-A curve signals the onset of the transition between the so-called liquid-expanded (LE) and liquid-condensed (LC) phases. Though observed and intensively studied in many similar systems, the nature of the LE-LC transition is still controversial. Experimental data are almost exclusively limited to π-A measurements and for their interpretation various assumptions about the molecular orientation are made but have never been checked experimentally. Figure 3 shows the results of our orientational measurements for PDA at 25°C as obtained from the SHG data. In the LE phase, θ rapidly increases with increasing density N_S until the LE-LC transition is reached, whereafter it changes more slowly and linearly with N_S. Here, θ refers to the angle between the surface normal and the polar C-OH bond, which appears to make the dominant contribution to the SHG from this molecule. Intuitively, one expects this polar bond to align normal to the water (and hence the molecular chains would tilt away from the surface normal) as indeed is observed at lower densities where θ approaches 0°. When N_S increases, the steric interaction between the hydrocarbon chains of neighboring molecules tends to align them towards, and hence forcing the C-OH orientation away from, the surface normal. At $N_S = 3.1 \times 10^{14}$ cm^{-2} a phase transition to an oriented liquid occurs. By measuring θ just below the LE-LC transition (in the LE phase) we found $\theta = 45° \pm 3°$ for

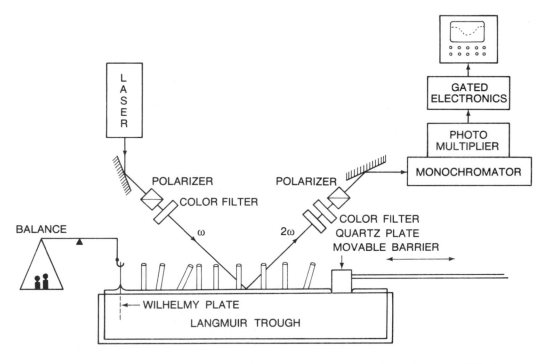

Figure 1. Schematic picture of the experimental apparatus to observe SHG from monolayers on a water surface.

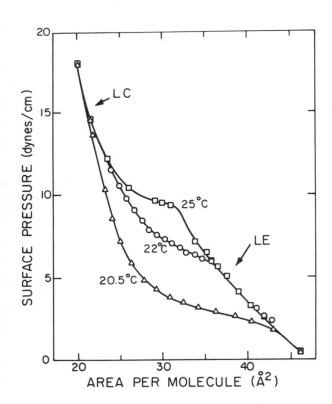

Figure 2. Surface pressure of PDA as a function of the area per molecule on a water surface of pH = 2 for various temperatures.

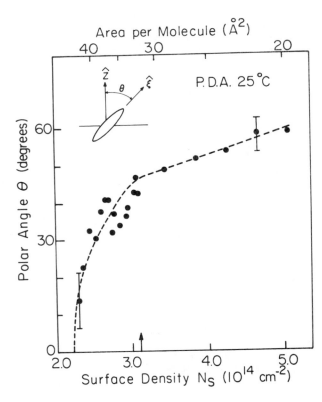

Figure 3. Tilt angle θ between the molecular axis and the surface normal as a function of the surface density of PDA on water at 25°C. The dashed line is an extrapolation through the data points.

all temperatures, though the transition point itself is very temperature dependent (see Fig. 2).

To measure $\overleftrightarrow{\alpha}^{(2)}$ of molecules, one usually relies on SHG measurements from powders[16] or dc electric field induced SHG (EFISH) from molecules in a liquid.[17] The problem with the first method is that one has to know the crystal structure and powder size distribution. With EFISH one actually measures an effective third-order nonlinear susceptibility, and to obtain $\overleftrightarrow{\alpha}^{(2)}$, a knowledge of solvent-solute interactions and local fields is essential. We recently introduced an alternative method by measuring SHG from an orientationally ordered monolayer on a water surface.[11] This works well for amphiphilic molecules, i.e. molecules with one hydrophobic and one hydrophilic part, and has the advantage of an easily controllable surface density of molecules and a low background. We have applied the technique on a number of cyanobiphenyl molecules $C_nH_{2n+1}(C_6H_4)_2CN$ (nCB, n = 8,9,10,12), several derivatives of this cyanobiphenyl structure, and a number of fatty acids $C_nH_{2n+1}COOH$ (nFA, n = 14,17,22). All the molecules used formed stable monolayers which were monitored by the SHG signal and the surface tension measurements and additionally checked by solubility tests.

Table I summarizes the obtained values for $\alpha^{(2)}_{\xi\xi\xi}$ and $\chi^{(2)}_{m,yzy}$. It shows that the $\alpha^{(2)}$ values of the biphenyl molecules are quite high [$\alpha^{(2)} \sim 2.5 \times 10^{-29}$ esu for 2-methyl-4-nitroaniline (MNA) at 1.06 μ [18]]. This high value results from the asymmetry induced in the π-electron system of the biphenyl rings by the presence of the CN group on one side and the hydrocarbon chain on the other side. The length of the latter has no effect on $\alpha^{(2)}$. However, the CN end group, which is a better electron acceptor than COOH, adds considerably to $\alpha^{(2)}$ as evidenced by the much smaller value of $\alpha^{(2)}$ for $C_8H_{17}(C_6H_4)_2COOH$ (8BCA). This is also consistent with the results for the fatty acids, where we find $\alpha^{(2)} \sim 10^{-31}$ esu, practically independent of chain length, indicating that $\alpha^{(2)}$ is dominated by the carboxylic acid group. Replacing a phenyl ring by a pyrimidine ring also causes a significant decrease in $\alpha^{(2)}$ as shown by the result for $C_7H_{15}(C_4N_2H_2)C_6H_4CN$ (7CPP). This presumably arises from an interruption of electron delocalization in the pyrimidine ring.[19] Adding another phenyl ring as in $C_5H_{11}(C_6H_4)_3CN$ (5CT) surprisingly leads to a decrease in $\alpha^{(2)}$ as well, and not to an increase as one might have expected as a result of a larger delocalized π-system.[20] We believe that this decrease may result from a nonplanar arrangement of the three phenyl rings in 5CT.

Table I. Second-order nonlinear polarizability of a number of organic molecules as determined by SHG from a monolayer using 532 nm excitation. The $\chi^{(2)}_{m,yzy}$ for the monolayer is given for a molecular density of $N_s = 3 \times 10^{14}$ cm^{-2}. θ is the angle between the molecular ξ axis and the surface normal.

Molecule		$\chi^{(2)}_{m,yzy}$ (10^{-16} esu) at $N_s = 3 \times 10^{14}$ cm^{-2}	$\theta(°)$	$\alpha^{(2)}_{\xi\xi\xi}$ (10^{-30} esu)
8CB	$C_8H_{17}(C_6H_4)_2CN$	11	71	25
9CB	$C_9H_{19}(C_6H_4)_2CN$	11	71	25
10CB	$C_{10}H_{21}(C_6H_4)_2CN$	11	71	25
12CB	$C_{12}H_{25}(C_6H_4)_2CN$	11	71	25
14FA	$C_{14}H_{29}COOH$	0.05	50	0.08
17FA	$C_{17}H_{35}COOH$	0.04	50	0.07
22FA	$C_{22}H_{45}COOH$	0.04	50	0.07
8BCA	$C_8H_{17}(C_6H_4)_2COOH$	2.8	63	6
7CPP	$C_7H_{15}(C_4N_2H_2)C_6H_4CN$	1.9	79	8
5CT	$C_5H_{11}(C_6H_4)_3CN$	3.5 (532 nm) 6 (586 nm)	60	7.5 (532 nm) 13 (586 nm)

As well as being employed in microlithography[2] and microelectronics,[3] monolayer polymerizations are of fundamental interest since their reactivity and kinetics may be studied under controllable and variable conditions of molecular separation and orientation.[21] We have used SHG to study the polymerization of two long chain monomers: vinyl stearate (VS) and octadecyl methacrylate (ODMA) both spread as a monolayer at a water/air interface. Although these materials do not have large second order nonlinearities the SHG signal can still be used to follow the extent and kinetics of polymerization undergone by the monomer, without disturbance or destruction of the monolayer film.

Table II shows the SHG results for pure water, for a water surface covered with a monomer monolayer, and for a water surface covered with a monolayer of commercially available bulk polymerized sample of the corresponding polymer.[14] The observed SHG intensities are clearly different for the various cases. Furthermore, after irradiating the monomer monolayers for two hours with a weak UV lamp under nitrogen atmosphere the SHG signals became very similar to those of the authentic polymer monolayers, indicating an almost complete UV initiated polymerization (in the absence of UV radiation no change was

Table II. Relative intensities and polarization of second harmonic generation from a water surface covered with various monolayers.

System	Relative SHG Intensity[a]	Polarization Ratio[b]
Water only	100	2
Water + VS monolayer (27 $Å^2$/molecule)	260	1.5
Water + poly VS monolayer[c] (27 $Å^2$/monomeric unit)	170	0.5
Water + VS monolayer after UV irradiation	180	
Water + ODMA monolayer (26 $Å^2$/molecule)	370	
Water + poly ODMA monolayer[c] (26 $Å^2$/monomeric unit)	220	0.5
Water + ODMA monolayer after UV irradiation	250	

[a] The total output SHG signal generated using an input 532 nm laser field polarized at 45° to the plane of incidence.
[b] The ratio of s-polarized to p-polarized SHG output.
[c] Bulk polymerized polymer spread on water.

induced by either the probe laser or the ambient thermal conditions).

The observed decrease in SH intensity upon polymerization can be understood as follows: the second order optical nonlinearity mainly arises from chemical bonds in which the electron distributions are more readily distorted by optical excitations.[22] In VS and ODMA the π electrons in the double bonds are likely to dominate the nonlinearity. Since the polymerization breaks a carbon-carbon double bond, $\alpha^{(2)}$ will decrease.

In order to follow the kinetics of polymerization we also made SHG measurements during UV irradiation.[14] As shown in Fig. 4 the SHG intensity decreases continuously during the reaction. Unfortunately, due to the low values of $\alpha^{(2)}$ and the relatively small changes in SHG intensity during polymerization our measurements were not accurate enough to unequivocably distinguish between first and second order polymerization kinetics. For the case of poly VS and poly ODMA, with only one C=O bond per unit, analysis of the SHG polarization showed that this C=O bond was perpendicular to the water surface. This is in agreement with both theoretical predictions[23] and infrared analysis of monolayers and multilayers which had been transferred onto various substrates.[24-26].

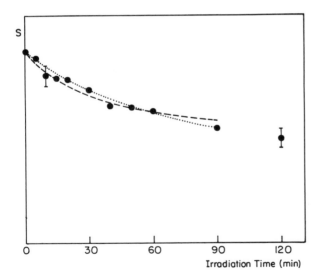

In conclusion, we have shown that SHG is a very sensitive and versatile probe to study molecular monolayers at an air-water interface. As examples we have studied molecular orientations and phase transitions, polymerization reactions and molecular nonlinear optical coefficients.

Figure 4. The relative SHG intensity (S) is plotted against irradiation time for UV polymerization of ODMA. The data (o) can be fitted satisfactorily by both first order (...) and second order (---) kinetics (see the text). All experimental data points have the same uncertainty, although error bars for most points have been omitted for clarity.

3. ACKNOWLEDGEMENTS

The authors have enjoyed collaboration with M. W. Kim and S. G. Grubb (Exxon, Annandale) on some of the projects described in this paper. GB acknowledges the financial support of a Chaim Weizmann Postdoctoral Fellowship. This work was supported by the Director, Office of Energy Research, Office of Basic Energy Sciences, Materials Sciences Division of the U.S. Department of Energy under Contract No. DE-AC03-76SF00098.

4. REFERENCES

*Permanent address: Department of Structural Chemistry, Weizmann Institute of Science, Rehovot 76100 Israel.

1. G. M. Bell, L. L. Coombs, and L. J. Dunne, Chem. Rev. 81, 15 (1981).
2. A. Barraud, C. Rosilio, and R. Ruaudel-Teixier, Thin Solid Films 68, 91 (1980).
3. G. L. Larkins, E. D. Thompson, M. J. Deen, C. W. Burkhart, and J. B. Lando, IEEE Trans. Magn. 19, 980 (1983).
4. J. Zyss, J. Mol. Elect. 1, 25 (1985).
5. J. F. Rabolt, F. C. Burns, N. E. Schlotter, and J. D. Swalen, J. Chem. Phys. 78, 946 (1983).
6. M. Seul, P. Eisenberger, and H. M. McConnell, Proc. Natl. Acad. Sci. USA 80, 5795 (1983).
7. Y. R. Shen, J. Vac. Sci. Technol. B3, 1464 (1985), and references therein.
8. Y. R. Shen, The Principles of Nonlinear Optics (J. Wiley, New York, 1984).
9. T. F. Heinz, H. W. K. Tom, and Y. R. Shen, Phys. Rev. A 28, 1883 (1983).
10. Th. Rasing, Y. R. Shen, M. W. Kim, P. Valint, Jr., and J. Bock, Phys. Rev. A 31, 537 (1985).
11. Th. Rasing, G. Berkovic, Y. R. Shen, S. G. Grubb, and M. W. Kim, Chem. Phys. Lett. 130, 1 (1986).
12. G. Berkovic, Th. Rasing, and Y. R. Shen, to appear in J. Opt. Soc. Amer. B.
13. Th. Rasing, Y. R. Shen, M. W. Kim, and S. Grubb, Phys. Rev. Lett. 55, 2903 (1985).
14. G. Berkovic, Th. Rasing, and Y. R. Shen, J. Chem. Phys. 85, 7374 (1986).
15. R. J. Priestley, ed., CRC Handbook of Lasers (Chemical Rubber Co., Cleveland, 1971), p.497.
16. A. F. Garito, K. D. Singer, and C. C. Teng, in Nonlinear Optical Properties of Organic and Polymeric Materials, ed. D. J. Williams (Amer. Chem. Soc., Washington, 1983), pp.1-26.
17. G. R. Meredith, in Nonlinear Optical Properties of Organic and Polymeric Materials, ed. D. J. Williams (Amer. Chem. Soc., Washington, 1983), pp.27-56.
18. B. F. Levine, C. G. Bethea, C. D. Thurmond, R. T. Lynch, and J. L. Bernstein, J. Appl. Phys. 50, 2523 (1979).
19. J. F. Nicoud and R. J. Twieg, in Nonlinear Optical Properties of Organic Molecules and Crystals, eds. D. S. Chemla and J. Zyss (Academic Press, Orlando, 1987), p.227.
20. A. Dulcic, C. Flytzanis, C. L. Tang, D. Pepin, M. Fetizon, and Y. Hoppilliard, J. Chem. Phys. 74, 1559 (1981).
21. S. A. Letts, T. Fort, Jr., and J. B. Lando, J. Colloid. Int. Sci. 56, 64 (1976).
22. D. J. Williams, Angew. Chem. Int. Ed. 23, 690 (1984).
23. D. Naegele and H. Ringsdorf, J. Polym. Sci. Polym. Chem. Ed. 15, 2821 (1977).
24. V. Enkelmann and J. B. Lando, J. Polym. Sci. Polym. Chem. Ed. 15, 1843 (1977).
25. K. Fukuda and T. Shiozawa, Thin Solid Films 68, 55 (1980).
26. S. J. Mumby, J. D. Swalen, and J. F. Rabolt, Macromolecules 19, 1054 (1986).

New Hemicyanine Dye-substituted Polyethers as Nonlinear Optical Materials

R. C. Hall and G. A. Lindsay
Chemistry Division, Naval Weapons Center, China Lake, California

S. T. Kowel
Department of Electrical and Computer Engineering, University of California, Davis

L. M. Hayden
Department of Physics, University of California, Davis

B. L. Anderson, B. G. Higgins, P. Stroeve and M. P. Srinivasan
Department of Chemical Engineering, University of California, Davis

Abstract

Highly oriented dye-containing polymer films are a promising class of nonlinear optical materials. We have prepared a poly(epicholorohydrin) derivative suitable for fabrication of noncentrosymmetric Langmuir-Blodgett (L/B) multilayer films. The synthesis, L/B deposition, and second harmonic generation measurements of these polymer films are reported.

Introduction

Organic nonlinear optical materials are becoming an increasingly popular area of research. Excellent literature reviews on the subject have been published,[1-3] and many corporate, university, and government laboratories are actively involved in development of organic nonlinear optical materials. Advantages over inorganics include large optical nonlinearities, high damage thresholds, and fast response times.

The development of processable polymeric nonlinear optical materials would eliminate problems associated with the growth of large, high quality single crystals. Incorporating the noncentrosymmetry required for second order nonlinear optical effects into polymer films has been approached by using molecularly doped and poled films[4,5] and oriented liquid crystalline polymers.[6] We have chosen the L/B technique for fabrication of noncentrosymmetric polymer films.

The L/B method is well suited for making highly ordered thin films of organic materials. This method requires use of amphiphilic molecules having a water soluble head group and a hydrophobic tail group such as a long chain fatty acid. The amphiphiles are spread on the water surface of a L/B trough. The hydrophilic head group sits on the water surface, while the hydrophobic alkyl tails orient themselves away from the water. By moving a barrier across the water surface and compressing the molecules, an organized monolayer is formed. A substrate is dipped through the monolayer, transferring it to the substrate. Subsequent dipping of this coated substrate through new monolayers builds a multilayered film. Surfactant polymers having a hydrophilic backbone and hydrophobic substituents have been used to form L/B films.[7-10]

Langmuir-Blodgett films of non-polymeric organic dyes have exhibited second order nonlinear optical effects,[11,12] and polydiacetylene L/B films have been extensively investigated[13-15] as third order nonlinear optical materials. Dye-polymer mixtures have shown second harmonic generation in L/B films.[16,17] By covalently incorporating nonlinear optical chromophores into the hydrophobic portions of L/B-type polymers, dye-polymer immiscibility can be avoided and second order nonlinear optical effects may be obtained. Monomers containing groups with large molecular hyperpolarizabilities such as 4-nitroanilines, 4-amino-4'-nitrostilbenes, and stilbazolium hemicyanines are often unreactive towards radical or ionic polymerization or give poor conversion to polymer.[18,19] The chemical attachment of polarizable dye groups to preformed polymers, however, avoids these problems.

We have prepared hemicyanine-substituted poly(epichlorohydrin) (PECH) of the structure shown in Figure 1. These polymers are specifically designed for L/B film formation in that they contain hydrophilic backbones and hydrophobic pendant groups.

Experimental

The synthesis of hemicyanine-substituted PECH, shown in Scheme I, is outlined as follows. PECH 4000 Molecular weight was obtained from 3M Co., and purified by reprecipitation. Base-catalyzed ether formation by reaction of the chloromethyl groups along the PECH backbone with 4-hydroxybenzaldehyde resulted in a methoxybenzaldehyde-substituted polyether. Comparing ratios of areas under proton NMR peaks showed substitution along the polymer backbone to range from about 30-50% for the several times this reaction was carried out (consistent with literature report on alkylation of phenols with PECH[20]). Condensation of the aldehyde-substituted polymer with long chain (C_{12}, C_{16}, C_{18}) alkyl-picolinium bromides yielded hemicyanine-substituted PECH[16]. Proton NMR showed all of the aldehyde sites had reacted, giving 30-50% dye-substitution along the backbone. These polymers are chloroform soluble, deep red colored glasses. The chromophores absorb visible light wth a lambda maximum of 390 nm and an absorption edge near 500 nm (10^{-4} M in chloroform), making them transparent to doubled Nd^{+3}-YAG laser light at 532 nm.

The Langmuir-Blodgett film formation was carried out as reported earlier[16] on a Joyce-Loebl LPB4 trough. Monolayers of hemicyanine-substituted PECH were spread from chloroform onto pure glass-distilled water. The polymers were deposited onto two back-to-back clean glass slides on the upstroke with average deposition ratios of about 0.95 (for five layers). Deposition stroke speed was 0.1 mm/sec. The subphase surface was cleaned and behenic acid (C_{22}) was deposited on the downstroke, resulting in alternating layers of PECH-hemicyanine and behenic acid. This alternating method prevented Y-type deposition of the polymer and resulted in formation of noncentrosymmetric multilayers.

Second harmonic generation was measured as described elsewhere[21] by passing Nd^{+3}-YAG laser light (1064 nm) through mono- and multilayer films on a glass slide and detecting the second harmonic (532 nm) light with a photomultipier tube.

Results and Discussion

Of several dye-substituted polymers prepared, only the polymer with C_{12} aliphatic chains and 47% dye-substitution has been well characterized. The behavior of the C_{16} and C_{18} chain dye-substituted polymers will be described at a later date.[22]

Figure 2 shows the surface pressure vs. surface area (pi-A) curve for the PECH-hemicyanine-C_{12} 1 at 22° C. During monolayer compression the pi-A curve exhibits a plateau between 20 and 30 dynes/cm, then shows a steep slope until the monolayer collapses near 45 dynes/cm. This plateau may be due to the positively charged group at the top of the hemicyanine dye. At pressures below 20 dynes/cm the dyes could lie down on the subphase surface to solvate the charged species, resulting in a "log-jam" of horizontal chromophores in the compressed monolayer. Above 20 dynes/cm the dyes may stand up more vertically relative to the subphase. A similar plateau has been observed in the pi-A curves of non-polymeric amphiphilic hemicyanines.[16] Recently we have prepared a polymer with the hemicyanine chromophore inverted relative to PECH-hemicyanine-C_{12} so that the charged group is closer to the polymer backbone and the pi-A curve has no plateau.[22]

The area per chromophore for 1 at 40 dynes/cm is about 25 $Å^2$, and the area per polymer molecule at 40 dynes/cm is about 500 $Å^2$. It is presently not clear what the unsubstituted portions of the polymer are doing during compression of the monolayer.

The increase in relative second harmonic intensity with additional layers of 1 is shown in Figure 3. The enhancement of SHG intensity is roughly quadratic (within error bars) for the first three layers of polymer. These polymer layers are interleaved with layers of behenic acid. Preliminary experiments show second harmonic enhancement to be less than quadratic for additional polymer layers after the first three. Work is currently under way to improve enhancement in thicker films.

The thermal behavior of 1 was investigated by differential scanning calorimetry (DSC), thermal gravimetric analysis (TGA), and by heating on a hot stage under a polarizing microscope. A glass transition temperature (Tg°) of about 200° C was observed by DSC. Visual observation as well as DSC and TGA showed an onset of thermal decomposition near 250° C.

No evidence of liquid crystalline behavior was found for 1, though the possibility was recognized that the unsubstituted portions of PECH backbone may act as spacers for the mesogenic chromophores.[9,23]

Conclusion

Hemicyanine-substituted PECHs are transparent glassy polymers suitable for fabrication of noncentrosymmetric L/B films. PECH-hemicyanines are promising materials for electric field-induced orientation because of the high dye-concentrations obtainable through chemical attachment to the polymer. Their high solubility in chloroform and high Tg°s make the PECH-hemicyanines potential candidates for integration with electronic devices.

Acknowledgements

This work was primarily supported by the Americal Society for Engineering Education and the Office of Naval Technology.

References

1. Garito, A. F., Singer, K. D.; Laser Focus, 18, 59 (1982).
2. Williams, D. J. ed.; ACS Symp. Ser. 233 (1983).
3. Williams, D. J.; Angew. Chem. Intl. Eng., 23, 690 (1984).
4. Sohn, J. E., Singer, K. D., Lalama, S. J., Kuzyk, M. G.; Proc. ACS Div. PMSE, 55, 532 (1986)
5. Pantelis, P., Hill, J. R., Davies, G. J.; Presented at ACS Div. Phys. Chem., 193rd Annual Meeting, Denver (1987).
6. Stamatoff, J. B., Buckley, A., Calundann, G., Choe, E. W., Demartino, R., Khanarian, G., Leslie, T., Nelson, G., Stuetz, D., Teng, C. C., Yoon, H. N.; Proc. SPIE, 682, 85 (1986).
7. Winter, C. S., Tregold, R. H., Vickers, A. J., Khoshdel, E., Hodge, P.; Thin Solid Films, 134, 49 (1985).
8. Tregold, R. H.; Presented at NSF Workshop on Molecular Engineering of Ultrathin Polymer Films, Davis, California (1987).
9. Laschewsky, A., Ringsdorf, H., Schmidt, G., Schneider, J.; J. Am. Chem. Soc., 109, 788 (1987).
10. Berkovic, G., Rasing, Th., Shen, Y. R.; J. Chem. Phys. 85(12) 7374 (1986).
11. Girling, I. R., Cade, N. A., Kolinsky, P. V., Earls, J. D., Cross, G. H., Peterson, I. R.; Thin Solid Films, 132, 101 (1985).
12. Neal, D. B., Petty, M. C., Roberts, G. G., Ahmad, M. M., Feast, W. J., Girling, I. R., Cade, N. A., Kolinsky, P. V., Peterson, I. R.; Presented at IEEE Intl. Symp. on Applications of Ferroelectrics (1986).
13. Carter, G. M., Chen, Y. J., Georger, JR., J., Hryniewicz, J., Rooney, M., Rubner, M. F., Samuelson, L. A., Sandman, D. J., Thakur, M., Tripathy, K.; Mol. Cryst. Liq. Cryst., 106, 259 (1984).
14. Yoshioka, Y., Nakahara, H., Fukuda, K.; Thin Solid Films, 133, 11 (1985).
15. Prasad, P. N.; Presented at ACS Div. Phys. Chem., 193rd Annual Meeting, Denver (1987).
16. Stroeve, P., Srinivasan, M. P., Higgins, B. G., Kowel, S. T.; Thin Solid Films, 146, 209 (1987).
17. Kowel, S. T., Ye, L., Zhang, Y., Hayden, L. M.; Opt. Eng. 26(2) 107 (1987).
18. Le Barney, P., Ravaux, G., Dubois, J. C., Parneix, J. P., Njeumo, R., Legrand, C., Levelut, A. M.; Proc. SPIE, 682, 56 (1986).
19. Griffin, A. C., Bhatti, A. M., Hung, R. S. L.; Proc. SPIE, 682, 65 (1986).
20. Percec, V., Pugh, C.; ACS Polymer Preprints, 27(2), 36 (1986).
21. Hayden, L. M., Kowel, S. T., Srinivasan, M. P.; Opt. Comm., 61(5), 351 (1987).
22. Lindsay, G. A., Hall, R. C., Anderson, B. L., Higgins, B. G., Stroeve, P., Kowel, S. T.; To be presented at Mat. Res. Soc. Boston meeting, Fall 1987.
23. Zentel, R., Reckert, G.; Makromol. Chem., 187, 1915 (1986).

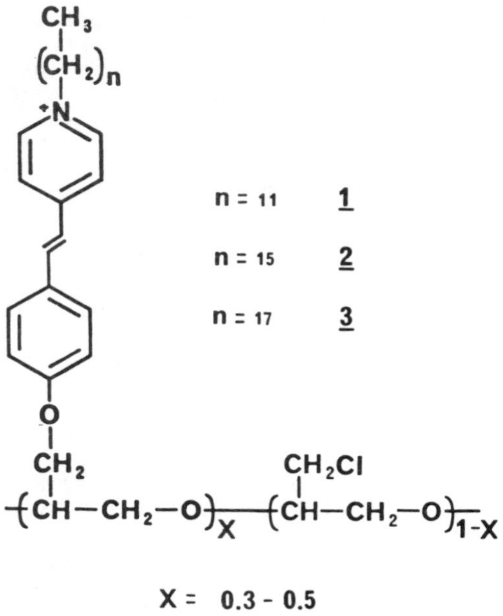

Figure 1. Hemicyanine-substituted PECH.

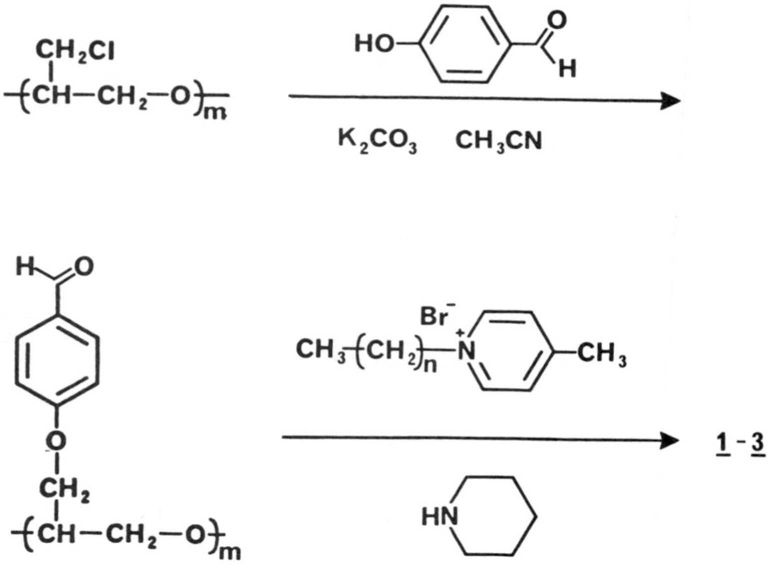

Scheme I. Synthesis of Hemicyanine-substitued PECH.

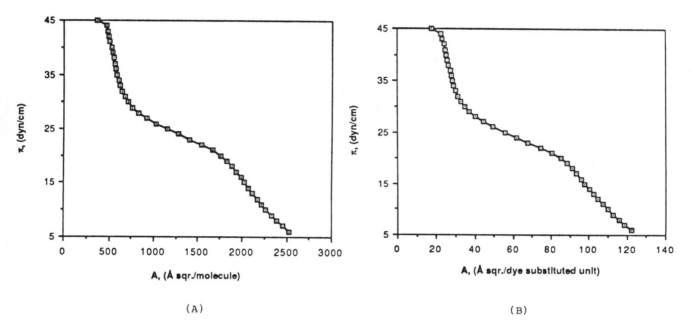

Figure 2. Surface Pressure vs. Surface Area curve for PECH-Hemicyanine-C12 $\underline{1}$ shown as (A) area per polymer molecule and (B) area per hemicyanine-substituted unit.

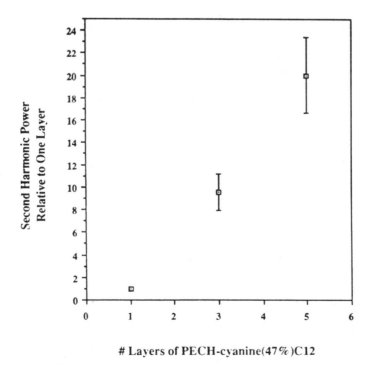

Layers of PECH-cyanine(47%)C12

Figure 3. Second harmonic intensity vs. number of layers for PECH-Hemicyanine-C$_{12}$.

Shortcomings of Molecular Models of Macroscopic Nonlinear Polarization

Gerald R. Meredith

E. I. du Pont de Nemours and Co., Central Research and Development Department
Experimental Station 356, Wilmington, Delaware, 19898

Abstract

A formal relation between molecular polarization and macroscopic nonlinear dielectric susceptibilities is derived. The importance of local field effects is reiterated. Application to crystalline and random media are discussed. To test validity of averaging procedures and semiempirical local-field models some optical third-harmonic generation experiments in binary solutions are described.

Introduction

The dielectric behavior of ensembles of molecules is usually described with language that assumes the persistence of molecular identity and properties in condensed phases.[1,2] As a working model this is reasonable since the intermolecular interactions, even of the most severe type (such as hydrogen bonding or even intermolecular charge transfer), are at least an order of magnitude lower than the energies associated with the intramolecular bonding. Thus the ensemble dipolar response to external electric fields can be envisioned to be an accumulation of contributions from intramolecular electron-cloud polarization, atomic framework deformation, molecular rotation, electrodeformation, electrophoretic effects, etc. In this paper this model will be pursued formally and experimentally in the realm of nonlinear dielectric response to electric fields in optical waves.

The dependence of molecular properties, including linear and nonlinear polarizabilities, on environment is generally recognized, but its study and impact are completely interwoven with the subject of "local fields" or mutual molecular polarization. Accurate handling of the latter is essential if these variations of molecular properties are to be understood and if the dielectric properties of condensed phase materials are to be truly understood (and made the object of precise molecular engineering). There is, of course, the objection that this viewpoint disregards the fact that in condensed-phase materials, the material excitations which are fundamentally responsible for the dielectric response are collective. Consequently, polarization processes are not correctly described in such a simplified picture. However, since the extent to which collective excitations are important in describing nonlinear ensemble polarization has not been well studied (at least in part because their consequences do not differ enormously from those of the simpler model) and since the above-described molecular picture is widely used, it is worthwhile despite this objection to review and further investigate the implications of the simpler model.

In the next section a formal summation and partitioning of molecular contributions to the ensemble polarization is summarized. Infinite summations appear which are grouped into commonly cited quantities. Due to the mutual polarization of molecules through "local fields", two important processes occur: 1) "cascading" of lower-order nonlinear polarizabilities into the higher-order nonlinear susceptibilities, and 2) "pull down" of higher-order polarizabilities into the lower-order susceptibilities. These infinite summations are functionally worthless, so averaging processes are described in following sections. In highly ordered media, i.e. crystals, simple expressions result. However those derived here differ from those in the widely quoted work of Zyss and Oudar[3] by inclusion of the tensorial nature of "local field" effects. In less ordered media the averaging process leads to a dilemma: how should the averaging process be performed in these complicated expressions? Even for such straightforward processes as second or third harmonic generation quite different results are anticipated with the more correct averaging compared to often quoted formulas. Finally, some third harmonic generation experiments in binary solution liquids are described which were undertaken to see evidence of these deviations.

A Formal Description of the Molecular Origin of Macrosopic Polarization

In this section a straightforward derivation of the nonlinear dielectric constants of a molecular material is summarized. With the intention to use these expressions where no zero-frequency perturbations of the ensemble result, a frozen-gas model is adopted. Should this not be the case, the final expressions could be assumed to be the zero-perturbation expressions for the ensemble and additional "ensemble distribution perturbations" could be added by a straightforward Taylor expansion.

The starting point is the recognition that the microscopic field which polarizes molecule κ is the sum of the Maxwellian electric field plus contributions from the dipole moments on all of the other molecules λ in the ensemble,

$$\mathbf{E}_m{}^\kappa = \mathbf{E} + \Sigma_\lambda \, \mathbf{L}^{\kappa\lambda} \cdot \mathbf{p}^\lambda \quad . \tag{1}$$

If one assumes point dipoles, one can adopt the relationship

$$\mathbf{L}^{\kappa\lambda} = (3\mathbf{r}^{\kappa\lambda} \, \mathbf{r}^{\kappa\lambda} - |\mathbf{r}^{\kappa\lambda}|^2) \, / \, |\mathbf{r}^{\kappa\lambda}|^5 \quad . \tag{2}$$

However, it is well known that the point dipole approximation is not good for larger, anisotropic molecules which are densely arranged. Nevertheless, in the spirit of single polarizability tensors and a uniform field acting on each molecule, it is assumed there are some $\mathbf{L}^{\kappa\lambda}$ which describe this relationship between $\mathbf{p}^\lambda$ and its contribution of field at molecule κ. If one defines the tensors $\mathbf{M}^{\kappa\lambda}$ and their "inverses" $\mathbf{R}^{\lambda\mu}$,

$$\mathbf{M}^{\kappa\lambda} = \delta_{\kappa\lambda} \, \mathbf{U} - \alpha^\kappa \cdot \mathbf{L}^{\kappa\lambda} \quad , \tag{3a}$$

$$\Sigma_\lambda \, \mathbf{M}^{\kappa\lambda} \cdot \mathbf{R}^{\lambda\mu} = \delta_{\kappa\mu} \, \mathbf{U} \quad , \tag{3b}$$

one may obtain a tensor,

$$\mathbf{N}^\kappa = \Sigma_\lambda \, (\mathbf{R}^{\lambda\kappa})^\mathsf{T} \quad , \tag{4}$$

which will be identified below as the "local field tensor".

The dipole moment on each molecule is the sum of the permanent, linearly induced and the nonlinearly induced moments,

$$\mathbf{p}^\kappa = \mu^\kappa + \alpha^\kappa \cdot \mathbf{E}_m{}^\kappa + \mathbf{p}^{(nl)\kappa} \tag{5}$$

$$\mathbf{p}^{(nl)\kappa} = \beta^{\kappa} \cdot \cdot \, \mathbf{E}_m{}^\kappa \mathbf{E}_m{}^\kappa + \gamma^{\kappa} \cdot \cdot \cdot \, \mathbf{E}_m{}^\kappa \mathbf{E}_m{}^\kappa \mathbf{E}_m{}^\kappa \ldots \quad . \tag{6}$$

Simple substitution and rearrangement yields

$$\Sigma_\kappa \, \mathbf{p}^\kappa = \Sigma_\kappa \{ \, (\mathbf{N}^\kappa)^\mathsf{T} \cdot \mu^\kappa + \alpha^\kappa \cdot \mathbf{N}^\kappa \cdot \mathbf{E} + (\mathbf{N}^\kappa)^\mathsf{T} \cdot [\beta^{\kappa} \cdot \cdot \, \mathbf{E}_m{}^\kappa \mathbf{E}_m{}^\kappa + \ldots] \, \} \quad . \tag{7}$$

One sees how with only linear polarizability $\mathbf{N}^\kappa$ appears formally to convert the Maxwellian to the microscopic field. It is defined only in terms of the linear polarizability, being not a bad approximation in general since nonlinear polarizability is small in comparison except in very intense fields. It is convenient to collect some terms into a new tensor

$$Q^{\kappa\mu} = \Sigma_\lambda \, (R^{\lambda\kappa})^T \cdot L^{\lambda\mu} \ , \tag{8}$$

which functions as a "dressed dipole transfer tensor". Two subsequent relationships, which demonstrate this functionality,

$$E_m{}^\kappa = N^\kappa \cdot E + \Sigma_\mu \, Q^{\kappa\mu} \cdot (\mu^\mu + p^{(nl)\mu}) \quad , \tag{9}$$

$$(N^\kappa)^T \cdot \mu^\kappa = \Sigma_\lambda \, (\, \delta_{\kappa\lambda} \, U - \alpha^\lambda \cdot Q^{\kappa\mu} \,) \cdot \mu^\kappa \quad , \tag{10}$$

also suggest the definition of another tensor,

$$D^\kappa = \Sigma_\lambda \, Q^{\kappa\lambda} \cdot \mu^\lambda \tag{11}$$

which looks like the induced electric field at molecule κ due to all of the permanent dipoles - and their distributions - in the ensemble.

With the above expressions and definitions one can collect terms by powers in the Maxwellian field and identify them as the bulk constants.

$$P_0 = V^{-1} \, \Sigma_\kappa \, (N^\kappa)^T \cdot \{ \, \mu^\kappa + \beta^\kappa \cdot\cdot \, D^\kappa D^\kappa + \gamma^\kappa \cdot\cdot\cdot D^\kappa D^\kappa D^\kappa$$
$$+ \, 2 \, \beta^\kappa \cdot\cdot \, D^\kappa \, (\, \Sigma_\lambda \, Q^{\kappa\lambda} \cdot \beta^\lambda \cdot\cdot \, D^\lambda D^\lambda \,) + ... \, \} \tag{12}$$

$$\chi^{(1)} = V^{-1} \, \Sigma_\kappa \, \{ \, (\alpha^\kappa)^T \cdot N^\kappa$$
$$+ (N^\kappa)^T \cdot [2 \, \beta^\kappa \cdot\cdot \, D^\kappa N^\kappa + 3 \, \gamma^\kappa \cdot\cdot\cdot D^\kappa D^\kappa N^\kappa$$
$$+ 2 \, \beta^\kappa \cdot\cdot \, (\, \Sigma_\lambda \, Q^{\kappa\lambda} \cdot \beta^\lambda \cdot\cdot \, D^\lambda D^\lambda \,) \, N^\kappa + ... \,]\} \tag{13}$$

$$\chi^{(2)} = V^{-1} \, \Sigma_\kappa \, (N^\kappa)^T \cdot \{ \, \beta^\kappa \cdot\cdot \, N^\kappa N^\kappa + 3 \, \gamma^\kappa \cdot\cdot\cdot D^\kappa N^\kappa N^\kappa$$
$$+ 2 \, \beta^\kappa \cdot\cdot \, D^\kappa \, (\, \Sigma_\lambda \, Q^{\kappa\lambda} \cdot \beta^\lambda \cdot\cdot \, N^\lambda N^\lambda \,)$$
$$+ 4 \, \beta^\kappa \cdot\cdot \, N^\kappa \, (\, \Sigma_\lambda \, Q^{\kappa\lambda} \cdot \beta^\lambda \cdot\cdot \, D^\lambda N^\lambda \,)$$
$$+ 6 \, \delta^\kappa \cdot\cdot\cdot D^\kappa D^\kappa N^\kappa N^\kappa + ... \tag{14}$$

$$\chi^{(3)} = V^{-1} \, \Sigma_\kappa \, (N^\kappa)^T \cdot \{ \, \gamma^\kappa \cdot\cdot\cdot \, N^\kappa N^\kappa N^\kappa$$
$$+ 2 \, \beta^\kappa \cdot\cdot \, N^\kappa \, (\, \Sigma_\lambda \, Q^{\kappa\lambda} \cdot \beta^\lambda \cdot\cdot \, N^\lambda N^\lambda \,)$$
$$+ 4 \, \delta^\kappa \cdot\cdot\cdot D^\kappa N^\kappa N^\kappa N^\kappa + ... \, \} \tag{15}$$

etc.

The large number of terms arises because the microscopic fields occur 1) due to E and N^κ, 2) due to μ^μ, and 3) due to $p^{(nl)\mu}$ (see Eq. 9). If only the first mehanism were active, there would be a one-to-one relationship between orders of microscopic nonlinear polarizability and macroscopic nonlinear susceptibilities. However, the other two sources of microscopic fields respectively "pull down" higher-order molecular nonlinearities into lower-order susceptibilities and "cascade" lower-order molecular nonlinearities into higher-order susceptibilities. These two mechanisms are almost never included in descriptions of molecular behavior, only the first terms of Eqs. 12-15 usually being assumed. Considering the discovery that highly nonlinear molecular species exist, one can see the importance of their inclusion, if mutual polarization phenomena are important.

In Eqs. 12-15 multiplicative numeric factors occur because terms are generated werein the ordering of factors are interchanged. Since the presentation above does not specifically label frequency, the tensors contain all frequencies. If one wishes to move to frequency labelling, such as $\beta(-2\omega, \omega, \omega)$, in the expressions, one must further count the additional identical terms which would occur due to that procedure.

These expressions are not very useful in their infinite summation form. Given the complexity of their calculation, it is realistic to acknowledge the form and adopt ensemble averaging processes, perhaps using them in a semiempirical manner.

Importance of Mutual Polarization

As can be seen with simple estimates, <u>the local field correction is extremely important</u>. If one adopts the widely used Lorentz-Lorenz model, $N = (n^2+2)/3$. If n varies between 1.5 and 2, N varies between ~4/3 and 2, and disregarding averaging details, the triple product which occurs in $\chi^{(2)}$ would be in the range of 2.8 - 8! The quadruple product in $\chi^{(3)}$ would be even larger, and so on. Clearly, the <u>local field factors potentially contribute more to the macroscopic nonlinear susceptibilities than the simple sum of intrinsic molecular nonlinear polarizabilities. If their properties are not accurately known and if they are not accurately treated, then a) accurate knowledge of molecules inferred from bulk susceptibilities and b) accurate knowledge of the origins of bulk susceptibilities from molecular hyperpolarizabilities cannot be obtained.</u>

Given the magnitude of local-field effects one might question studies where their uncertainty are severe limits. Surfaces are an obvious example. Less obvious would be computer calculations of the nonlinearity of molecular aggregates. When the result exceeds or deviates from the sum of molecular nonlinearities, how does one separate the mutual polarization effects and does the result, from a limited surrounding of polarizable material, have any bearing on the effects in crystals or liquids?

Local-Field Effects in Crystals

To utilize Eqs. 12-15 one must perform some ensemble averaging. In perfect crystals this is straightforward. Unit cells are equivalent by translation; each primitive unit cell contains Z nontranslationally equivalent molecules, or equivalently, there are Z sublattices of translationally equivalent molecules; the nontranslationally pure operations of the crystal's space group relate these, there often being only one environment or type molecule in the lattice. Below this is assumed to be the case. When there is more than one type of molecular site or environment, simple extension of the formalism may be done.

There being only one crystal molecular environment, disregarding the effects of small fluctuations due to strain fields, thermal motions, surfaces or impurities, the averaging process can be reduced to averaging over molecules of the unit cell and using the unit cell volume for V. For example consider the often used approximation of $\chi^{(2)}$ which is simply the first term of Eq. 14,

$$\chi^{(2)} = V^{-1} \Sigma_q \, (N^q)^T \cdot \, \beta^q \cdot\cdot \, N^q N^q \quad . \tag{16}$$

Because of the interchange symmetry between the molecules of the unit cell, the β and N are the same matrices in the Z molecular reference frames. However, they are not the same in the unit cell or crystal lattice reference frame in which the summation must be performed. The process of rotating tensors and adding has been described by Zyss and Oudar[3] disregarding the tensorial nature of N^q. In the ansatz of their Eq. 1.1, they pull the local fields out of the summation and simply sum the β^q. This is a questionable step, but, regardless, the above expression can be recast to utilize their results for the orientational averaged per molecule hyperpolarizability, b_{IJK}, of their Eq. 1.3.

In a molecular (site) reference frame the tensors will have forms (i.e., the distribution of elements in matrices) β and N which are identical. If A^q is the real orthogonal matrix which transforms the elements of a vector $\boldsymbol{\nu}$ expressed in site q reference frame into V^q expressed in the unit cell reference frame,

$$V^q = \mathbf{A}^q \cdot \boldsymbol{\nu} \quad , \tag{17}$$

then Eq. 16 becomes,

$$\chi^{(2)} = V^{-1} \Sigma_q (\mathbf{A}^q \cdot \boldsymbol{\mathcal{N}} \cdot \mathbf{A}^{q-1})^T \cdot$$
$$(\mathbf{A}^q \cdot \beta \cdot\cdot \mathbf{A}^{q-1} \mathbf{A}^{q-1}) \cdot\cdot (\mathbf{A}^q \cdot \boldsymbol{\mathcal{N}} \cdot \mathbf{A}^{q-1}) (\mathbf{A}^q \cdot \boldsymbol{\mathcal{N}} \cdot \mathbf{A}^{q-1}) \tag{18a}$$

$$= V^{-1} \Sigma_q \mathbf{A}^q \cdot \{ (\boldsymbol{\mathcal{N}})^T \cdot \beta \cdot\cdot \boldsymbol{\mathcal{N}} \boldsymbol{\mathcal{N}} \} \cdot\cdot \mathbf{A}^{q-1} \mathbf{A}^{q-1} \quad . \tag{18b}$$

There are two important aspects of $\boldsymbol{\mathcal{N}}$: amplification and rotation of the field. Unfortunately, one cannot in general decompose $\boldsymbol{\mathcal{N}}$ into a product of a scalar and rotation matrix. If so,

$$\boldsymbol{\mathcal{N}} = f \, \mathcal{A} \quad , \tag{19}$$

$$\chi^{(2)} = V^{-1} f f f \Sigma_q \mathbf{A}^q \cdot \beta' \cdot\cdot \mathbf{A}^{q-1} \mathbf{A}^{q-1} \tag{20a}$$

$$\beta' = (\mathcal{A})^T \cdot \beta \cdot\cdot \mathcal{A} \mathcal{A} \quad . \tag{20b}$$

The ansatz Eq. 1.1 of Zyss and Oudar has a form similar to Eq. 20a (after adding frequency labels) if the distinction between β' and β is ignored. However, the amplification of the local field is a function of direction which does not in general fit the form of the Zyss and Oudar ansatz. (For a fuller discussion see the Appendix.) A major aspect of that work dealt with "optimizing molecular orientations", and should be rethought in view of the simplicity of their local field treatment.

This problem of microscopic fields from densely packed anisotropic molecules has complicated (nonlinear) spectroscopies in molecular crystals, fostering various theoretical approaches meant to provide useful insights.[4-6] The use of symmetry has been common in molecular crystal spectroscopy. At the least, the point tensor model may be simplified by using the full site point group to establish the form of β and $\boldsymbol{\mathcal{N}}$ matrices. There is ample spectroscopic information that the crystal forces are impressed up molecules and that the reduction of molecular symmetry to the site symmetry occurs, thus changing molecular properties, even if one does not consider the problem of collective excitations.[7]

Local-Field Effects in Disordered Media

The treatment of mutual polarization in liquids is an old and long subject. Increasingly precise measurements and interpretation of dielectric constants, refractive index, Kerr effects and Raman/Rayleigh scattering have necessitated more detailed considerations of local-fields and correlations. Historically continuum cavity models, such as those of Lorentz, Lorenz, Debye and Onsager have been used, giving reasonably good agreement with linear properties.[1,2,8] Such models produce formulas which are close to more realistic statistically derived results. One can speculate on this good fortune and reasons for this occurrence. (Consider that first-order corrections to such models detrimentally affect their performance while often adding unknown parameters to the expressions.) Surprisingly, correlation effects in passing from the first term of Eq. 13 to the susceptibility of refractive index is not significant. Correlations not contained in averaging of the type $\langle a_{11} \cdot b \rangle_{11} \approx \langle a \rangle_{11} \cdot \langle b \rangle_{11}$ are not large (consider the microscopic interdependence and the effects of amplitude fluctuations). For nonlinear processes one must wonder whether the same good fortune applies.

Consider the averaging process necessary to pass from the first summed term of Eq. 15 to the often quoted formula for a neat liquid,

$$\chi^{(3)} = \rho N \; f \, f \, f \, f < \gamma >_{\text{orientation only}} \qquad .$$

(See references 9-11 for discussions and demonstrations of the effects in the second term.) Major approximations had to be made to get to this point. One must question whether all fluctuations and correlations of local-field and molecular polarizability magnitudes and directions are reasonably handled by this adaptation from the linear results. Simply questioning the effects of local-field amplitude fluctuation raises skepticism. It's easily shown that $<a^4> \gg <a>^4$ in nonconstant distributions. If the fluctuations or deviations among all the factors are totally uncorrelated, then certainly expressions such as above are justified. This may be more the case in Kerr experiments or to a lesser degree in dc-electric-field-induced second-harmonic generation where the fields at different frequencies affect different molecular degrees of freedom and to some extent decouple the fluctuations or deviations of environment. At any rate, it was considered that optical third-harmonic generation would be the most direct and critical test of the optical frequency field models since, except for a small dispersion enhancement at the harmonic, the local fields are identical.

A Test of Local-Field Models by Optical Third-Harmonic Generation

A very direct test would be the characterization of molecular nonlinearity in the vapor phase, then in condensed fluid phase. Such experiments are under consideration, but require extensive redesign of experimentation (due to low vapor pressures of interesting molecules and due to longer optical path requirements) and, more importantly, require that the calibration procedure be careful chosen to be the same in both experiments. The search for fluctuation effects was, then, indirect. It was supposed that differences in their effects would be observable in the behavior of susceptibility as a function of mole fraction in binary miscible liquids series, if the two liquids were substantially different in molecular properties.

Unfortunately, a second difficulty arises in the use of solutions: will both chemical species be assumed to experience the same (fluctuating) local field or not? Since local fields are in part the consequence of reaction fields due to the polarization on a molecule, if one type molecule has larger volume-normalized polarizability, shouldn't it experience a larger average local field? Singer and Garito[12] addressed this problem in dc-electric-field-induced second-harmonic generation by offering Lorentz-Lorenz (all molecules are treated equally) or Onsager-type optical local-field models. Unfortunately, an algebraic error in their derivations removed the distinction and they concluded there was no difference. This issue was considered to be a secondary question which might be partly illuminated by this study.

Solutions of nitrobenzene (Kodak triply distilled, electronics grade) and methanol (Burdick and Jackson spectroquality, distilled in glass) were prepared and characterized by previously published means.[9] Precision in the total process of determining THG susceptibilities was apparently ~2%. The susceptibilities were clearly seen not to be simply molecular-wise additive functions. Local fields must consequently be employed. On applying the single field Lorentz-Lorenz model, agreement was seen to within experimental error. Interestingly, in connection with the nonadditivity, on factoring the common local-field factor, additivity is restored and a linear behavior of the data was observed. The use of the more complex Onsager-type model also gave equally good performance, though the nonfactoring of the local-field prevents the additivity observation. It was expected that substantial deviations from these mean-field models would be observed in the midrange of mole fraction where the fluctuations and range of environments would be largest. We were unable to see a consistent trend in the data.

These experiments demonstrated the necessity of using local fields, but indicated that the effects of fluctuations and environmental variations are no larger in mixtures than neat liquids. Similar results were obtained with other pairs of chemicals combined into solutions.

Closing

The effects of mutual polarization have been discussed. With the increasing interest in nonlinear optical properties of organic materials, it is important to keep in mind the severe restrictions on molecular characterization and on bulk material predictions that they cause.

References

1. C. J. F. Bottcher, <u>Theory of Electric Polarization</u>, 2nd ed. revised by O. C. Van Belle, P. Bordewijk and A. Rip (Elsevier, New York, 1973) Vol. I
2. C. J. F. Bottcher and P. Bordewijk, <u>Theory of Electric Polarization</u> (Elsevier, New York, 1978) Vol. II
3. J. Zyss and J. L. Oudar, Phys. Rev. <u>A26</u>, 2028 (1982)
4. F. P. Chen, D. M. Hanson and D. Fox, J. Chem. Phys. <u>66</u>, 4954 (1977); <u>63</u>, 3878 (1975)
5. P. J. Bounds and R. W. Munn, Chem. Phys. <u>24</u>, 343 (1977); and later papers by R. W. Munn
6. R. M. Hochstrasser, C. M. Klimcak and G. R. Meredith, J. Chem. Phys. <u>70</u>, 870 (1979)
7. E. R. Bernstein and G. R. Meredith, J. Chem. Phys. <u>64</u>, 375 (1976)
8. K. Vedam, CRC Crit. Rev. Sol. St. Mat. Sci. <u>11</u>, 1 (1983); K Vedam and P. Limsuwan, J. Chem. Phys. <u>69</u>, 4762, 4772 (1980)
9. G. R. Meredith, B. Buchalter and C. Hanzlik, J. Chem. Phys. <u>78</u>, 1533, 1543 (1983)
10. G. R. Meredith, Chem. Phys. Lett. 92, 165 (1982)
11. G. R. Meredith and B. Buchalter, J. Chem. Phys. <u>78</u>, 1938 (1983)
12. K. D. Singer and A. F. Garito, J. Chem. Phys. <u>75</u>, 3572 (1981)

Appendix

The difference between the treatment of local fields in molecular crystals here and by Zyss and Oudar is in essence the difference between local-field tensors active at each molecular <u>site</u> or averaged to the <u>unit cell</u> level. One might argue that considering the physical situation, distributed polarizability exists, which has the effect of smoothing the anisotropic nature of $\mathbf{N}$, thus increasing the validity of the latter approximation. Nevertheless, the molecular picture formally requires the use of the site model.

Both site and unit-cell local-field models must describe all of the measurable linear and nonlinear susceptibilities. Using $\mathbf{n}$ to signify the unit-cell local-field tensor (in its most general form), and equating expressions of $\chi^{(1)}$ for arbitrary polarizability, leads to the relation

$$\mathbf{n} = Z^{-1} \sum_q (\mathbf{A}^q \cdot \mathbf{N} \cdot \mathbf{A}^{q-1}) = Z^{-1} \sum_q \mathbf{N}^q \quad , \tag{A1}$$

i.e., <u>$\mathbf{n}$ is the average of the site local-field tensors.</u> Considering the two model expressions of $\chi^{(2)}$,

$$\chi^{(2)} = V^{-1} \sum_q \mathbf{A}^q \cdot \{ (\mathbf{N})^T \cdot \beta \cdot\cdot \mathbf{N}\,\mathbf{N} \} \cdot\cdot \mathbf{A}^{q-1}\,\mathbf{A}^{q-1} \tag{A2a}$$

$$\chi^{(2)} = V^{-1} (\mathbf{n})^T \cdot \{ \sum_q \mathbf{A}^q \cdot \beta \cdot\cdot \mathbf{A}^{q-1}\,\mathbf{A}^{q-1} \} \cdot\cdot \mathbf{n}\,\mathbf{n} \quad , \tag{A2b}$$

and their generalizations to higher-order susceptibilities, and using Eq. A1, leads to a general requirement on $\mathbf{N}$ <u>if the unit-cell approximation is to be valid</u>.

$$\mathbf{N} \cdot \mathbf{A}^q = Z^{-1} \sum_s (\mathbf{A}^q \cdot \mathbf{A}^{s-1} \cdot \mathbf{N} \cdot \mathbf{A}^s) \tag{A3a}$$

Use of (site-to-site) interchange operations[4,7] and their coordinate-vector matrices, $\mathbf{A}^{s \to q} = \mathbf{A}^q \cdot \mathbf{A}^{s-1}$, which convert between the q-site and r-site coordinate systems, allows simplification of this requirement to

$$\mathbf{N} = Z^{-1} \sum_s (\mathbf{A}^{s \to q} \cdot \mathbf{N} \cdot \mathbf{A}^{q \to s}) \quad . \tag{A3b}$$

Eq. A3b states approximately the same thing as Eq. A1, that is, for the unit-cell local-field approximation to hold the local-field tensor must be the average of the site local-field tensors expressed in the same coordinate system (here a site reference frame). But this is a restriction now on $\mathbf{N}$, not simply on $\mathbf{n}$, an additional constraint brought on by use of the unit-cell approximation in the nonlinear susceptibilities.

There can be up to 9 independent real components of $\mathbf{N}$ in the site local-field model (which may be reduced by symmetry in the event of site point groups higher than C_1). However, the unit-cell approximation eliminates part of $\mathbf{N}$ which do not have the full symmetry of the unit cell, reducing the number of independent elements and restricting its form (as can be seen in Eq. A3b). It is the loss of elements in $\mathbf{N}$ which brings into question the use of the Zyss and Oudar approach.

Incoherent Light Application to the Measurement of Vibrational and Electronic Dephasing Time

Takayoshi Kobayashi, Toshiaki Hattori, and Akira Terasaki

Department of Physics, Faculty of Science, University of Tokyo,
7-3-1 Hongo, Bunkyo-ku, Tokyo 113, Japan

Abstract

The convenient method for the short-time measurements using temporally incoherent light instead of short pulses was applied to electronic- and vibrational-dephasing-time measurement. Electronic dephasing in a polydiacetylene film and vibrational dephasing in dimethylsulfoxide liquid were observed with femtosecond time resolution.

1. Introduction

Static optical properties of condensed matter had been extensively studied and the dynamic behaviors started to attract increasing number of scientists when pulsed lasers became available. The development of picosecond and femtosecond spectroscopies in the last two decades has enabled to bring much information of the relaxation in condensed phase with improved time resolution. Since the picosecond-light-pulse generation was demonstrated using passively mode-locked ruby laser in 1965,[1] continuous efforts to generate shorter pulses have been made, and optical pulses as short as 8 fs were obtained[2] by the method of pulse compression of the output from a group-velocity-dispersion-compensated colliding-pulse mode-locked laser. Very recently the shortest pulse of 6 fs was obtained by the compression method using prism pair and grating pair for the compensation of both second- and third-order group velocity dispersions.[3] Time-resolved coherent and conventional spectroscopies have been applied to several systems including semiconductors and dye molecules using ultrashort light pulses with pulse width between a few tens and a hundred femtoseconds. However, there are several difficulties and disadvantages in the study of the ultrafast phenomena using such short pulses: (i) Lasers for the generation of ultrashort pulses applicable to spectroscopies are expensive and complicated. (ii) The wavelengths of pulsed laser with pulse width shorter than a few hundred femtoseconds are usually limited in the region around 615-625 nm because of the limitation in the appropriate combination of saturable absorber and gain medium, and the tunable range of each laser is generally narrow (usually less than 10 nm). (iii) It is difficult to avoid broadening of an ultrashort pulse due to linear and/or nonlinear dispersion in optical components because of its broad power spectrum.

Recently a new spectroscopic technique with incoherent light utilizing transient coherent optical effects has been proposed. In the transient coherent spectroscopy for the studies of the ultrafast dynamics, the time resolution is expected to be determined by the correlation time instead of the pulse duration. According to this principle, extremely high-time-resolution experiment can be performed much more easily by using temporally incoherent light with a short enough correlation time. The applicability of this principle to ultrafast process studies has been verified for the electronic-dephasing-time measurement by degenerate four-wave mixing (DFWM) spectroscopy.[4-8]

In Chapter 2, we describe the study of the electronic dephasing in a polydiacetylene (poly-3BCMU) film measured by spatial-parametric-mixing type DFWM (or sometimes called forward DFWM).[9] Dephasing time in polydiacetylene films was resolved for the first time. The experiment was performed using incoherent light at two wavelengths. The measured dephasing times, 30 fs at 648 nm and 130 fs at 582 nm, correspond to excitons in polymer chains associated with the electronic states having different conjugation lengths.

Though transient DFWM spectroscopy, either using coherent short pulses[10] or incoherent cw light sources as mentioned above, is powerful for the study of dynamic processes in condensed matter, it cannot be applied to optically forbidden transition and the available wavelength range is highly limited. Dephasing of Raman active vibrational modes in molecules can be observed by so-called transient coherent Raman spectroscopy such as CARS (coherent anti-Stokes Raman scattering) and CSRS (coherent Stokes Raman scattering),[11-13] where a pair of picosecond pulses excites a vibrational system coherently and a second pulse of the higher (or lower) frequency probes the coherence of the system after a certain delay time. The information about the dephasing dynamics of the system can be obtained by the delay-time dependence of the coherent Raman scattering intensity.

We studied theoretically a possible application of the above-mentioned principle that the correlation time determines the resolution time, to transient coherent Raman spectroscopy.

Theoretical derivation of the delay-time dependence of the coherent Raman intensity, and the experimental demonstration of the determination of the dephasing time of the 2915-cm^{-1} mode in dimethylsulfoxide are described in Chapter 3.[14]

2. Electronic dephasing in a polydiacetylene film measured by degenerate four-wave mixing

There has been much interest in the dynamical properties of the excited states of polydiacetylenes (PDAs), which form a group of compounds of the typical quasi-one-dimensional systems. They have been studied experimentally by time-resolved absorption, reflection, and emission spectroscopy.[15-20] The lifetime of soluble PDA (poly-3KAU) in the lowest excited singlet state was found to be 9±3 ps in aqueous solution[15] and that of PTS to be about 2 ps in crystalline phase.[16]

Information about the dephasing dynamics of PDAs is of great importance not only for the elucidation of the electronic properties of excited states and the mechanism of the optical nonlinearity but also for various applications such as optical switching and optical signal processing. DFWM was applied to the dephasing time measurement,[21-23] but dephasing times have not been resolved so far. Dennis et al.[21] observed DFWM from two PDA (2d and 2j) solutions with 180 ps pulses, but the response times were much shorter than their resolution time. Carter et al.[22] observed DFWM from a PDA (PTS) crystal with 6 ps pulses, and the response time was again not able to be resolved. Rao et al.[23] performed similar measurements on a PDA (poly-4BCMU) film with 500 fs pulses, but they could not resolve the dephasing time either.

In the present study, we applied DFWM with incoherent light to the measurement of the dephasing times in a cast film of a PDA, poly[4,6-decadiyne-1,10-diol bis((n-butoxycarbonyl)-methyl) urethane], which is abbreviated as poly-3BCMU. By detecting signals diffracted in two directions simultaneously, we could resolve a dephasing time as short as 30 fs. We measured phase-relaxation times of the sample at two wavelengths, 648 nm and 582 nm, and found that the dephasing of the exciton in a chain of the polymer with a longer conjugation length (at 648 nm) is four times faster than that in a chain with a shorter conjugation length (at 582 nm). This result is consistent with the difference in the fluorescence efficiencies between the rod-like and coil-like forms.

2.1 Experimental

The experimental apparatus used for the dephasing time measurement by the spatial-parametric-mixing type DFWM is shown in Figure 1. The incoherent light source was a broad-band dye laser pumped by a N$_2$ laser. In order to obtain very broad-band laser light, the dye laser cavity without any tuning element was constructed with a highly-reflecting aluminum mirror and a glass plate as an output mirror. Rhodamine 6G and rhodamine 640 were the laser dyes. The peak wavelengths and the band widths (FWHMs) of the power spectra were 582 nm and 7.7 nm for the rhodamine 6G laser, respectively, and 648 nm and 8.7 nm for the rhodamine 640 laser, respectively.

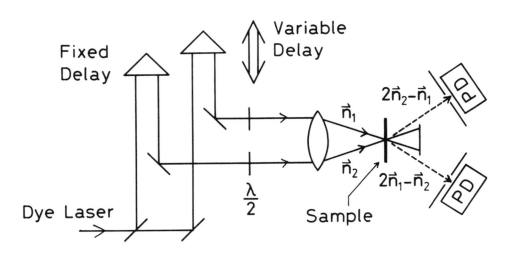

Figure 1. Experimental setup for spatial-parametric-mixing type degenerate four-wave mixing measurement. PD and $\lambda/2$ stand for a photodiode and a half-wave plate (Fresnel rhomb), respectively. The vectors, $\mathbf{n}_1$ and $\mathbf{n}_2$, are the unit vectors representing the directions of the two excitation beams.

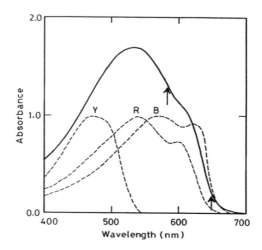

Figure 2. Absorption spectrum of the film of poly-3BCMU (solid line) with those of three forms in solution from Reference 25 (broken lines). B, R, and Y denote blue, red, and yellow forms, respectively. Two exciting wavelengths utilized in the present study, 648 nm and 582 nm, are indicated by arrows.

The broad-band dye laser light was linearly polarized and divided into two beams, n_1 and n_2, and n_2 was delayed with respect to n_1 by a variable delay line. The polarization planes of the two beams could be rotated independently by the use of half-wave optics composed of two Fresnel rhombs. The pulse energies of the 10 ns dye laser were about $3\,\mu$J at the sample position, and the beam diameters of the focused areas on the sample were $40\,\mu$m. Degenerate four-wave mixing signals diffracted in two directions, $2n_2-n_1$ and $2n_1-n_2$, were detected simultaneously by photodiodes, as shown in Figure 1.[8] Output signals of the photodiodes were processed with sample-and-hold circuits and an A/D converter. The sample was a film of poly-3BCMU on a glass plate cast from chloroform solution. The absorption spectrum of the film is shown in Figure 2.

2.2 Results and discussion

Figure 3 shows the data which were obtained with the two excitation beams under parallel polarization condition. Background signal due to scattering of the excitation beams was subtracted from the data. No peak shift or tail was observed. The signal with parallel polarizations was about thirty times more intense than that with perpendicular polarizations, and it can be attributed almost exclusively to a thermal grating, which is generated only when mutual coherence between the two excitation beams exists.[10]

The contribution of the thermal grating to the DFWM signals can be eliminated under perpendicular polarization, and hence we can obtain electronic DFWM signals,[8] which are shown in Figure 4 after the background scattering intensity being subtracted. The peaks of the signal intensities of the two directions are shifted from each other. Here we define a peak separation as the distance in the delay time between the two signal peaks of the two directions. The peak separations were 30 and 90 fs at 648 and 582 nm, respectively. The widths (FWHMs) of the signals were 100 fs for 648 nm and 130 fs for 582 nm, which give approximately the correlation time of the incoherent excitation light at each wavelength. There is no pronounced asymmetric tails which indicate dephasing times much longer than the correlation time. The tails seen in the data at 648 nm are the same as those seen in thermal grating signals, and they reflect the shape of the laser spectrum.

Under resonance excitation condition, the delay-time dependence of the signal intensities is expressed by the following equation:[7]

$$I(t_d) \propto \int_o^\infty dt \int_o^\infty dt' G(t'-t)G(t-t_d)G^*(t'-t_d)\exp[-2(t+t')/T_2].\qquad(1)$$

Here T_2 is the dephasing time, t_d is the delay time, and $G(t_d)$ is the autocorrelation function of the incoherent light field. When the dephasing time is much longer than the correlation time of the light field, the signal decays exponentially at the rate of $4/T_2$.

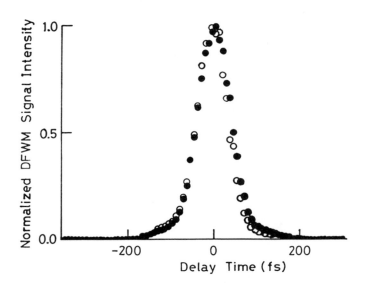

Figure 3.   DFWM signals obtained with the two excitation beams of polarizations parallel to each other.   The wavelength of the excitation light was 648 nm.   Open circles show the signal intensity diffracted in the direction $2n_1-n_2$, and closed circles show that in the direction $2n_2-n_1$.

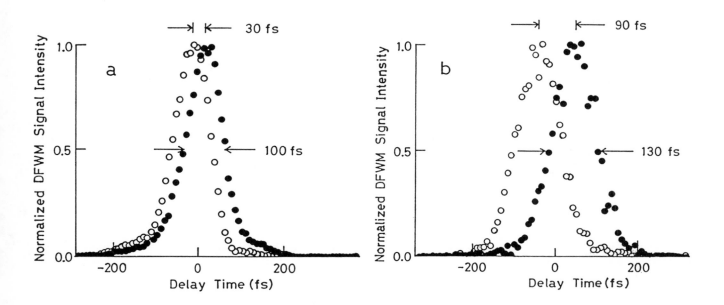

Figure 4.   DFWM signals with the two excitation beams of polarizations perpendicular to each other.   Open circles show the signal intensity diffracted in the direction $2n_1-n_2$, and closed circles show that in the direction $2n_2-n_1$.   The wavelengths of the excitation light are a) 648 nm and b) 582 nm.

On the other hand, when the dephasing time is comparable with or even slightly shorter than the correlation time, the signal shapes have no prominent tails. This is the case in the present study as presented later, but the dephasing time can also be obtained from the peak separation of the two signals diffracted in two directions.[8,10] Using Equation (1) and the observed peak separations, the dephasing times are calculated to be 30 fs at 648 nm and 130 fs at 582 nm by assuming Gaussian autocorrelation functions. These values change little even when other autocorrelation functions are assumed.

We would like to mention here the effect of the width of the inhomogeneous broadening on the magnitude of the peak separation. Equation (1) is correct only for extremely broad inhomogeneous widths compared with the homogeneous width. For narrower inhomogeneous width the peak separation becomes smaller, and there is no shift for completely homogeneous system. Since, in actual materials, the inhomogeneous width is finite, the dephasing time estimated by the observed peak separation may be shorter than the real value. In the present case, the absorption spectrum width of the sample is as broad as 4500 cm^{-1}, and the estimated reduction of the magnitude of the peak shift is about 4 %. Therefore, the effect of finite inhomogeneous width is negligible.

The present result that the dephasing time at a longer wavelength is shorter than that at a shorter wavelength is contrary to those obtained in previous studies.[8,10] In these studies concerned with laser dyes, the difference in the dephasing times was attributed to that in the intramolecular relaxation rate or that in the excess photon energy of the excitation light from the absorption edge.

The present sample is a polymeric film which has a quasi-one-dimensional structure and electronic properties different from ordinary dyes.[24,25] It is known that poly-3BCMU in solution has three forms with different absorption spectra. They are blue (B), red (R), and yellow (Y) forms.[24] The blue form is considered to have a polymer main chain with rod-like conformation, while R and Y forms have main chains with coil-like conformation. These three forms are realized depending on the temperature and the composition of the solvent. The shoulder of the absorbance of the sample in the present study at 620 nm (2.0 eV) corresponds to the B band, and the peak at 530 nm (2.3 eV) corresponds to the R band. They are attributed to the π-π^* exciton transition in each type of the polymer chain. Therefore, our sample is regarded as a mixture of a coil-like conformation (with shorter conjugation lengths of π-electron) and a rod-like conformation (with longer conjugation lengths), or a mixture of the polymer chains with continuously distributed conjugation lengths between these two extreme forms realized in solution.

In the present experiment, 648 nm is at the absorption edge of the rod-like form exciton, while 582 nm is on resonance with the exciton in polymer chains with shorter conjugation lengths. Therefore, from the present results on the dephasing times at the two wavelengths, it can be concluded that the exciton dephasing is several times faster in the rod-like form than in a chain with a shorter conjugation length.

This result may be explained by the following two mechanisms. One explanation is that excitons are more mobile and the phases of them are changed more often in longer conjugated chains than shorter ones, where excitons do not move over long distances. The other explanation is that exciton levels lie more closely in longer conjugated chains than in shorter conjugated chains, and therefore, dephasing due to multilevel excitation is faster.[10] We cannot determine which is the case only from the present data. An extended study at other wavelengths and temperatures is in progress.

It has been reported that the fluorescence intensities are suppressed when the solution of poly-3BCMU is converted from a yellow (coil-like) form to a blue (rod-like) form, and when the solution of PDA (poly-4BCMU) is converted from the coil form to the rod form.[24] It is also reported that only partially polymerized crystal of PDA (PTS) emits fluorescence.[26] Our results are consistent with the explanation[24,26] by which the changes in the fluorescence quantum efficiencies were attributed to the increase in the nonradiative decay rates with exciton delocalization.

3. Vibrational dephasing time measurement by coherent Stokes Raman scattering

We extended the method for the observation of electronic dephasing dynamics in picosecond to femtosecond region discussed in the previous section to the vibrational dephasing. A type of nondegenerate four-wave mixing using both coherent and incoherent light is utilized for the latter experiment. In the present paper the first theoretical study is presented and experimental description follows. It is based on a transient coherent Raman process with three beams (see Figure 5), two of which are incoherent light from a single broad-band laser, and a delay time between the two is variable. A third beam has a higher frequency than the incoherent light by a vibrational energy in a molecule of interest, and is coherent in the delay-time range of the measurement. The information about the coherence dynamics of

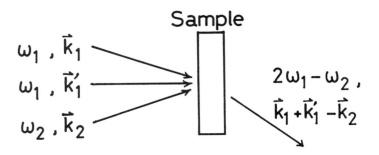

Figure 5. Schematic of CSRS experiment for vibrational dephasing measurement.

the vibrational transition can be obtained with a resolution time limited by the correlation time of the incoherent light from the delay-time dependence of coherent Stokes Raman scattering (CSRS) intensity. When the frequency of the third beam is lower than the others, coherent anti-Stokes Raman scattering (CARS) takes place. For the theoretical calculation, a three-level model of molecular system with homogeneous broadening is used, and the delay-time dependence of CSRS intensity was calculated. Dephasing dynamics of the 2915-cm^{-1} mode in dimethylsulfoxide was observed experimentally by the new method using a broad-band nanosecond dye laser, and the obtained dephasing time was found to agree well with that obtained with picosecond pulses. Raman echo experiment with incoherent light is also proposed in the last part of this chapter.

3.1 Description of a model and time dependence of signal intensity

All reported studies of transient spectroscopy with incoherent light, including photon echo and pump-probe spectroscopy, as far as we know, were concerned with only one broad band (about 100 cm^{-1}) within which exists the resonance frequency to the transition between two levels in systems. However, there exist various coherent transient phenomena where light beams of two or more different frequencies are concerned, and the time resolution of the transient coherent spectroscopies using these phenomena is also expected to be determined not by the duration of light pulses used but by the correlation time of the radiation field.

Theoretical expectation values of the coherent Raman signal intensity using incoherent light will be presented with a simple model in this section. Calculation of the time-dependent CSRS intensity will be presented for the correspondence with the experimental study described in the following section, although CARS is substantially the same.

For theoretical considerations of coherent Raman phenomena, a simple three-level system (see Figure 6) is usually taken as a model.[27,28] Two vibrational levels |1> and |2> belong to the ground electronic state, whereas level |3> belongs to an electronically excited state. In ordinary coherent Raman experiments, this system is placed in radiation field which consists of light beams of two frequencies, the difference between which is resonant with the transition between |1> and |2>. The higher frequency is denoted by ω_{AS} and the

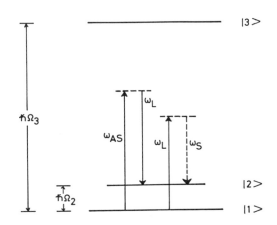

Figure 6. Energy diagram of the model system for vibrational dephasing experiment.

lower by ω_L. For the purpose of time-resolved measurement, a triple-beam (BOXCARS) configuration was applied,[29] where two beams of frequency ω_L are used, and three waves are mixed to generate a wave of frequency $2\omega_L - \omega_{AS}$.

The electric field is given as

$$E(\mathbf{r},t) = E_{AS}(t)\exp[i(\mathbf{k}_{AS}\mathbf{r}-\omega_{AS}t)] + E_{L1}(t)\exp[i(\mathbf{k}_{L1}\mathbf{r}-\omega_L t)] + E_{L2}(t)\exp[i(\mathbf{k}_{L2}\mathbf{r}-\omega_L t)]$$

$$+ \text{ c.c. } , \tag{2}$$

where c.c. stands for complex conjugates of the preceding terms and $E_{AS}(t)$, $E_{L1}(t)$, and $E_{L2}(t)$ are functions of t slowly varying compared to the optical frequencies. In BOXCARS experiments, light of frequency $\omega_S = 2\omega_L - \omega_{AS}$ and with wave vector $\mathbf{k}_S = \mathbf{k}_{L1}+\mathbf{k}_{L2}-\mathbf{k}_{AS}$ is detected.

The polarization with frequency ω_S and wave vector $\mathbf{k}_S$ is derived by a perturbation method under the following conditions assumed; (i) Light frequencies are tuned exactly to the vibrational energy. (ii) Both of the light of frequencies ω_{AS} and ω_L are off-resonance with electronic transitions. (iii) The broadening of the relevant energy level of the molecular system is homogeneous. Under these conditions, a third-order polarization $P^{(3)}(\mathbf{k}_S, \omega_S)$ is given by

$$P^{(3)}(\mathbf{k}_S, \omega_S) = C \exp[i(\mathbf{k}_S\mathbf{r}-\omega_S t)] \int_{-\infty}^{t} dt' \exp[-(t-t')/T_2]$$

$$\times [E_{L1}(t)E_{L2}(t')+E_{L2}(t)E_{L1}(t')]E_{AS}^{*}(t') , \tag{3}$$

where T_2 is the dephasing time of the vibrational transition, and C is a time-independent proportionality coefficient. This expression has two terms, in which E_{L1} and E_{L2} are exchanged with each other.

Two incoherent light beams of central frequency ω_L are obtained by splitting a beam from a broad-band dye laser in the present CSRS experiments using incoherent light. One of the two beams is delayed to the other with a delay time t_d, and the light of frequency ω_{AS} is assumed to be coherent in the delay time region of observation.

If the correlation time of the incoherent light is assumed to be much shorter than the dephasing time, and the stochastic property of the incoherent light to be expressed in terms of a Gaussian random process, a simple expression for the signal intensity can be derived as follows:

$$I(t_d) = 1 + G(t_d) + (2\tau_c/T_2)\exp(-2|t_d|/T_2) . \tag{4}$$

Here G(t) is the autocorrelation function of the incoherent light field amplitude normalized to unity at its peak, and the correlation time τ_c is defined by

$$\tau_c = \int_{0}^{\infty} G(t)dt . \tag{5}$$

The fraction of the third term in Equation (4) is approximately proportional to the ratio of the correlation time to the dephasing time. Therefore, for the dephasing time determination with high precision, incoherent light with an appropriate correlation time must be used, since low intensity signal cannot be distinguished from the background due to inevitable noises.

3.2 Experiment and results

CSRS signal by the symmetric CH-stretching vibration of dimethylsulfoxide (DMSO) with a wavenumber 2915 cm^{-1} was measured by the apparatus shown in Figure 7. Coherent light source was the second harmonic (532 nm) of a Q-switched Nd:YAG laser operated at 8 Hz. Since the correlation time of this light source is estimated to be about 30 ps, it can be regarded safely as coherent for the time period shorter than 10 ps. The main part (about 90% intensity) of this beam was split off and used to pump a broad-band dye laser. By changing the laser dye concentration, the oscillation wavelength was tuned to 630 nm which is resonant with the Raman mode of DMSO. The spectral width (FWHM) of the laser light was 7 nm, which corresponds to a correlation time of about 100 fs.

The dye laser beam was split into two with nearly equal intensity by a beam splitter. Both of them were made parallel again by the same beam splitter after passing through delay lines. The length of one of the delay lines was varied by a stepping motor. The three beams were focused in DMSO in a 10-mm sample cell by a lens of 20-cm focal length. To eliminate the scattered laser light, an iris with 5-mm diameter, a 10-cm monochromator, and color filters were placed between the sample and the photomultiplier.

The CSRS signal intensity is plotted, in Figure 8, as a function of the delay time in a semilogarithmic scale after the background is subtracted. From the slope of the trailing part, the vibrational dephasing time T_2 is found to be 1.4 ps in agreement with the value obtained with picosecond pulses.[12] The intensity ratio between the peak and the tail extrapolated to $t_d=0$ is about seven, which is also in good agreement with the theoretically expected ratio for the values of T_2 (1.4 ps) and the correlation time (100 fs) determined above.

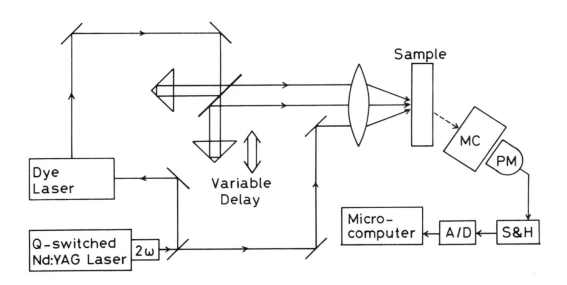

Figure 7. Experimental setup for CSRS measurement using incoherent and coherent light. MC is the monochromator; PM, the photomultiplier.

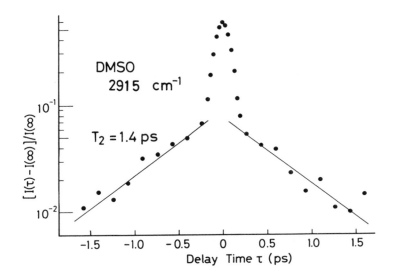

Figure 8. The delay-time dependence of the CSRS intensity normalized by the background intensity. The background intensity has been subtracted.

3.3 Proposal of Raman echo experiment

Since CSRS or CARS does not include a rephasing process, which takes place in photon echoes, the obtained value of the dephasing time has some ambiguities when the system is inhomogeneously broadened.[30,31] Raman echo[27] can provide the information about the dephasing dynamics without the ambiguities.[28] However, it is a higher-order (seventh-order) process, and the detection of the signal is very difficult. Raman echo experiments have been carried out in solids,[32] gases,[33,34] and liquid nitrogen,[35] but not yet in liquids at room temperature.

The following type of Raman echo experiment may be feasible with incoherent light (see Figure 9). The molecules are brought in coherent superposition of the ground state and a vibrationally excited state by light of two frequencies, ω_{AS} and ω_L at t=0. The difference between these frequencies is matched to the vibrational transition. Rephasing is introduced by light of the same pair of the frequencies at t=τ. The macroscopic coherence is recovered, if the vibrational transition is inhomogeneously broadened, at 2τ, and it is probed by the light of frequency ω_{AS}, and the emitted light of frequency $\omega_e = 2\omega_{AS} - \omega_L$ is detected. The two pulses of frequency ω_L in the figure are replaced in the actual experiment by two incoherent light beams with a delay of τ obtained from the same light source.

Since the probing pulse is not a short pulse but light with a constant intensity in the present experimental scheme, the emitted signal light may include a portion due to the free Raman susceptibility decay (the Raman analogue of the free induction decay). However, the dephasing time can be estimated fairly well by the echo decay rate and the background intensity even in this case. The expression for the signal intensity as a function of the delay time is completely the same as the one for the dephasing measurement by two-beam DFWM with incoherent light whether the system has either homogeneous or inhomogeneous broadening.[4] Of course, the incoherent light can be used as the probe pulse for the detection of only the echo signal. For this purpose a third incoherent beam which is delayed by τ with respect to the second one is used as the probe light, and the emitted light of frequency $2\omega_L - \omega_{AS}$ is detected.

4. Summary

We have reported here two types of ultrashort-time measurements using incoherent light, that is, studies of electronic and vibrational dephasing. This technique is effective in these studies and also in other measurements such as the determination of population relaxations of excited states and the Kerr effect dynamics. The potential applicability is quite extensive because of the simplicity and the tunability of the lasers used in this method.

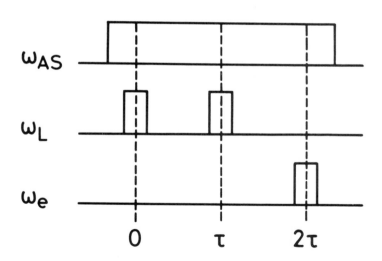

Figure 9. The temporal relationship among coherent light at ω_{AS} (long correlation time), incoherent light at ω_L (short correlation time), and the echo signal in the Raman echo experiment.

Acknowledgements

We wish to thank Professor T. Kotaka for providing us with poly-3BCMU. This work is supported partly by a Grant-in-Aid for Special Distinguished Research (56222005) from the Ministry of Education, Science and Culture of Japan.

References

1. Mocker, H. W., Collins, R. J., Appl. Phys. Lett., Vol. 7, pp. 270-273, 1965.
2. Knox, W. H., Fork, R. L., Downer, M. C., Stolen, R. H., Shank, C. V., Appl. Phys. Lett., Vol. 46, pp. 1120-1121, 1985.
3. Brito-Cruz, C. H., Fork, R. L., Shank, C. V., MD1 in Technical Digest of Conference on Lasers and Electro-optics, April 27-May 1, 1987, Baltimore, Maryland, USA.
4. Morita, N., Yajima, T., Phys. Rev. A, Vol. 30, pp. 2525-2536, 1984.
5. Asaka, S., Nakatsuka, H., Fujiwara, N., Matsuoka, M., Phys. Rev. A, Vol. 29, pp. 2286-2289, 1984.
6. Beach, R., Hartmann, S. R., Phys. Rev. Lett., Vol. 53, pp. 663-666, 1984.
7. Nakatsuka, H., Tomita, M., Fujiwara, M., Asaka, S., Opt. Commun., Vol. 52, pp. 150-152, 1984.
8. Fujiwara, M., Kuroda, R., Nakatsuka, H., J. Opt. Soc. Am. B, Vol. 2, pp. 1634-1639, 1985.
9. Hattori, T., Kobayashi, T., Chem. Phys. Lett., Vol. 133, pp. 230-234, 1987.
10. Weiner, A. M., Silvestri, S. De, Ippen, E. P., J. Opt. Soc. Am. B, Vol. 2, pp. 654-661, 1985.
11. Laubereau, A., Kaiser, W., Rev. Mod. Phys., Vol. 50, pp. 607-665, 1978.
12. George, S. M., Auwester, H., Harris, C. B., J. Chem. Phys., Vol. 73, pp. 5573-5583, 1980.
13. George, S. M., Harris, A. L., Berg, M., and Harris, C. B., J. Chem. Phys., Vol. 80, pp. 83-94, 1984.
14. Hattori, T., Terasaki, A., Kobayashi, T., Phys. Rev. A, Vol. 35, pp. 715-724, 1987.
15. Koshihara, S., Kobayashi, T., Uchiki, H., Kotaka, T., Ohnuma, H., Chem. Phys. Lett., Vol. 114, pp. 446-450, 1985.
16. Carter, G. M., Hryniewicz, J. V., Thakur, M. K., Chen, Y. J., Meyler, S. E., Appl. Phys. Lett., Vol. 49, pp. 998-1000, 1986.
17. Kobayashi, T., Iwai, J., Yoshizawa, M., Chem. Phys. Lett., Vol. 112, pp. 360-364, 1984.
18. Kobayashi, T., Ikeda, H., Tsuneyuki, S., Chem. Phys. Lett., Vol. 116, pp. 515-520, 1985.
19. Orenstein, J., Etemad, S., Baker, G. L., J. Phys. C, Vol. 17, pp. L297-300, 1984.
20. Robins, L., Orenstein, J., Superfine, R., Phys. Rev. Lett., Vol. 56, pp. 1850-1853, 1986.
21. Dennis, W. M., Blau, W., Bradley, D. J., Appl. Phys. Lett., Vol. 47, pp. 200-202, 1985.
22. Carter, G. M., Thakur, M. K., Chen, Y. J., Hryniewicz, J. V., Appl. Phys. Lett., Vol. 47, pp. 457-459, 1985.
23. Rao, D. N., Chopra, P., Ghoshal, S. K., Swiatkiewicz, J., Prasad, P. N., J. Chem. Phys., Vol. 84, pp. 7049-7050, 1986.
24. Kanetake, T., Tokura, Y., Koda, T., Kotaka, T., Ohnuma, H., J. Phys. Soc. Jpn., Vol. 54, pp. 4014-4026, 1985.
25. Chance, R. R., Patel, G. N., Witt, J. D., J. Chem. Phys., Vol. 71, pp. 206-211, 1979.
26. Sixl, H., Warta, R., Chem. Phys. Lett., Vol. 116, pp. 307-311, 1985.
27. Hartmann, S. R., IEEE. J. Quantum Electron., Vol. QE-4, pp. 802-807, 1968.
28. Loring, R. F., Mukamel, S., J. Chem. Phys., Vol. 83, pp. 2116-2128, 1985.
29. Eckbreth, A. C., Appl. Phys. Lett., Vol. 32, pp. 421-423, 1978.
30. Zinth, W., Polland, H.- J., Laubereau, A., Kaiser, W., Appl. Phys. B, Vol. 26, pp. 77-88, 1981.
31. George, S. M., Harris, C. B., Phys. Rev. A, Vol. 28, pp. 863-878, 1983.
32. Hu, P., Geschwind, S., Jedju, T. M., Phys. Rev. Lett., Vol. 37, pp. 1357-1360, 1976.
33. Leung, K. P., Mossberg, T. W., Hartmann, S. R., Phys. Rev. A, Vol. 25, pp. 3097-3101, 1982.
34. Brückner, V., Bente, E. A. J. M., Langelaar, J., Bebelaar, D., van Voorst, J. D. W., Opt. Commun., Vol. 51, pp. 49-52, 1984.
35. van Voorst, J. D. W., Brandt, D., van Hensbergen, B. L., in Technical Digest of Topical Meeting on Ultrafast Phenomena, 1986.

Optical Bistability in Dimer-Monomer Dye Systems

Shammai Speiser* and Frank L. Chisena

Allied-Signal Incorporated, Engineered Materials Sector
P.O. Box 1087R, Morristown, New Jersey 07960

ABSTRACT

Optical bistability has been observed in highly concentrated fluorescein dye solutions and in thin (~1 μm) doped polymeric films. At concentrations larger than 10^{-5} mole/ℓ dye dimers are formed. For fluorescein dye, the dimer-monomer equilibrium constant is 10^5 mole/ℓ so that most of the dye species are in the dimer form. At 480 nm the dimer absorption cross section is 10^{-18} cm²/molecule, while that for the dye monomer molecule is 7.6×10^{-17} cm²/molecule. Upon laser excitation dimers dissociate to form monomers thus providing a highly nonlinear laser induced absorption. This high nonlinear absorption coefficient can be utilized for optically bistable response of the dye system.

Optical bistability was observed by placing dye solutions or dye thin films inside a Fabry-Perot resonator and exciting it with 480 nm dye laser pulses of 10 ns duration. The effect is more pronounced in 10^{-4} mole/ℓ fluorescein than in 10^{-6} mole/ℓ fluorescein in which dimer formation is not that efficient.

1. INTRODUCTION

Optical bistability is characterized by two different light transmission states of an optical system for a given input light intensity.[1,2] In order to observe optical bistability a non-linear optical medium and an optical feedback are required. A non-linear absorbing medium in an optical resonator was the first configuration for which the existence of optical bistability was theoretically predicted by Szoke et al.,[3] by Siedel,[4] and by McCall.[5]

In order to analyze optical bistabilty for various systems, we have developed[6,7] the nonlinear complex eikonal approximation. Our goal was to set a standard mathematical treatment for the analysis of the propagation of light waves through nonlinear media. The lack of such a general treatment had complicated both the engineering modeling of optical systems incorporating nonlinear elements and better understanding of the related physical phenomena (since most of the other methods yield only numerical solutions for complicated cases).

The nonlinear eikonal approximation can be summarized in the following equation:

$$\phi(z) = (2\pi/\lambda_0) \int_0^z n[I(z')]dz' \tag{1}$$

where $\phi(z)$ is the complex accumulated phase, z - the distance of propagation in the medium, n - the non-linear complex index of refraction, I(z) - the local light intensity. This integral equation can be applied to a variety of nonlinear media (n(I)) and to general schemes (or boundary conditions) (I(z)). We have demonstrated[6,7] the applicability of the nonlinear eikonal approach for dispersive nonlinear resonators and we have obtained analytical results which were the first order approximation for the Fabry-Perot resonator case and the exact solution for the ring resonator. Recently we have extended the treatment to absorbing media[8] and in particular to molecular dye systems.[9] We were able to show that in addition to the obvious case of saturable-absorber optical bistability, one can expect to observe optical bistability due to excited singlet S_1 absorption.[9]

In many dye systems, however, complications due to dimer formation may destroy the desired nonlinear optical response. In this paper we treat such a system and demonstrate a novel type of optical bistability due to nonlinear absorption related to laser induced modification in the dye monomer-dimer equilibrium.

---

*On sabbatical leave. To whom correspondence should be addressed at Department of Chemistry, Technion-Israel Institute of Technology, Haifa 32000, Israel

2. EXPERIMENTAL

A Nd-YAG laser harmonic generated light (Quanta Ray DCR3) was used to excite a dye laser (Quanta Ray PDL-2). Dye laser light was used to excite dye solutions or dye polymer films placed in a Fabry-Perot resonator having mirror reflectivities of 90%. The input and output laser intensities are monitored by Hamamatsu 1188-06 PIN photodiodes and analyzed by a Tektronix 2430 digital oscilloscope interfaced to an IBM XT computer. Pump and probe experiments were performed using a 1% split off beam of the exciting laser, probing as right angle to the input beam. Laser induced fluorescence data were obtained by using a PTR monochromator and a Hamamatsu R928 photomultiplier.

Standard spin coating techniques were used to prepare thin fluorescein PMMA films.

3. RESULTS AND DISCUSSION

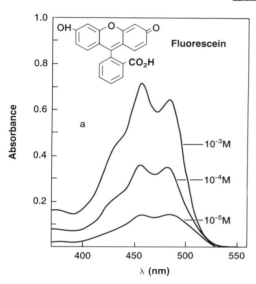
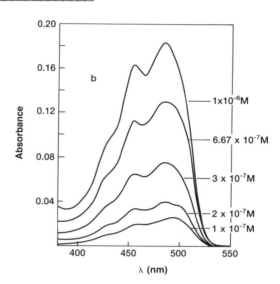

Figure 1. Asorption spectra for fluorescein solutions in ethanol (1 cm path length) (a) high concentration spectra (b) low concentration data.

Figure 1 shows the absorption spectrum of fluorescein at various concentrations. It is clear that Beer's law is not obeyed, mainly because of dimer formation.[10-12] Similar spectra were obtained for fluorescein PMMA thin films. The following kinetic scheme can be used to describe the dimer(A_2)-monomer(A) equilibrium:

$$A + A \underset{k_{DIS}}{\overset{k_{DIM}}{\rightleftharpoons}} A_2 \tag{2}$$

The effective absorption coefficient of the dye system is given by:

$$\beta = \varepsilon_A(C - 2X) + \varepsilon_{A_2}X \tag{3}$$

where C is the total molar concentration X is the equilibrium concentration of dimers and ε_A and ε_{A_2} are the molar excitation coefficients of the monomer and dimer species, respectively. From a simple equilibrium stoichiometry we obtain that

$$\beta = [(8KC + 1)^{1/2} + 1](\varepsilon_A - \varepsilon_{A_2}/2)/4K + C\varepsilon_{A_2}/2 \tag{4}$$

where $K = k_{DIM}/k_{DIS}$ is the association equilibrium constant.

At low concentrations, Beer's law is observed yielding $\varepsilon_A = 2 \times 10^4$ ℓ/mole cm. Eq. (4) can be fitted to the β vs C data of Fig. 1. The best fit curve is shown in Fig. 2

yielding $K = 10^5$ ℓ/mole and $\varepsilon_{A_2} = 260$ ℓ/mole cm, the deviation at high concentration is probably due to the presence of higher aggregates.

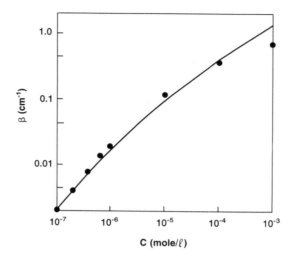

Figure 2.   Absorption coefficient β vs concentration for fluorescein, the solid line is the best fit curve to Eq (4).

Highly concentrated solutions of fluorescein which are highly dimerized might exhibit nonlinear absorption due to shift in the equilibrium of Eq (2) towards monomer formation. This was examined by utilizing the laser pump and probe technique. The absorption coefficient, $\alpha = 2.3\beta$, as a function pump intensity, I, is shown in Fig. 3, exhibiting a marked nonlinear absorption.

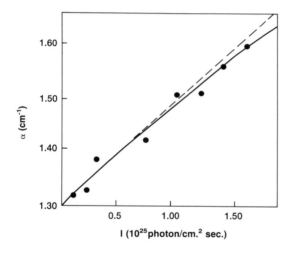

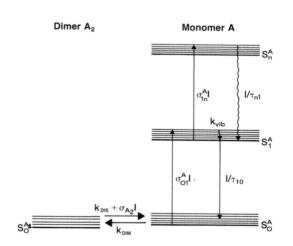

Figure 3.   Intensity dependent absorption coefficient, α for fluorescein 10^{-3} mole/ℓ ethanol solution at 480 nm. The solid line is the best fit curve to Eq. (7), the linear fit is to Eq. (7a).

Figure 4.   Schematic energy level diagram for a dimer-monomer dye system. All rate processes are slow compared to the vibrational relaxation rate, k_{vib}.

The increase in α as a function of pump intensity I can be analyzed by the following kinetic scheme of Fig. 4:

$$A_2(S_0) \underset{k_{DIM}}{\overset{k_{DIS}}{\rightleftharpoons}} 2A(S_0) \qquad\qquad (5a)$$

$$A2(S_0) \xrightarrow{\sigma_{A_2}I} 2A(S_0) \qquad\qquad (5b)$$

$$A(S_0) \underset{1/\tau_{10}}{\overset{\sigma_{01}^A I}{\rightleftharpoons}} A(S_1) \qquad\qquad (5c)$$

$$A(S_1) \underset{1/\tau_{n1}}{\overset{\sigma_{1n}^A I}{\rightleftharpoons}} A(S_n) \qquad (n > 1) \qquad\qquad (5d)$$

where S_0, S_1, and S_n denote ground, first excited and higher excited singlet states respectively . Dimers are dissociated by laser pumping at a rate $\sigma_{A_2}I$, where σ_{A_2} is the dimer absorption cross-section. From the absorption data of Fig. 3,[2] we calculate $\sigma_{A_2} = 3.83 \times 10^{-21}$ $\varepsilon_{A_2} = 10^{18}$ cm^2/molec. Monomers produced at a rate $\sigma_{A_2}I$ and are excited at a rate of $\sigma_{01}^A I$, where $\sigma_{01}^A = 7.66 \times 10^{-17}$ cm^2/molec. is the monomer absorption cross-section, τ_{10} is the monomer fluorescence life time and τ_{n1} is its S_n state lifetime.

Even for ns excitation, steady state conditions are reached[13] for fluorescein ($\tau_{10} = 4.0$ ns[13]). Under these conditions, we obtain that:

$$K(I) = A_2/A^2 = k_{DIM}/(k_{DIS} + \sigma_{A_2}I) = K/[1 + \sigma_{A_2}I/k_{DIS}] \qquad\qquad (6)$$

This analysis is valid for pump intensities that are too low to induce any photoquenching of the monomer fluorescence due to excitation of A (S_1) to higher singlet states at a rate $\sigma_{1n}^A I$ (σ_{1n}^A is the $S_1 \rightarrow S_n$ absorption cross-section, step (5d)).[13] The intensity dependent absorption coefficient is thus calculated from Eq. (4) to be:

$$\alpha(I) = \{[8K(I)N + 1]^{1/2} + 1\}(\sigma_{01}^A - \sigma_{A_2}/2)/4K(I) + N\sigma_{A_2}/2 \qquad\qquad (7)$$

where N is the molecular concentration (molecule/cm^3) $K(I)$ is in units of cm^3/molecule and I is in units of photon/cm^2 sec. At low intensities α approaches:

$$\alpha(I) = (\sigma_{01}^A - \sigma_{A_2}/2)[1 + 8KN)^{1/2} + 1](1 + \sigma_{A_2}I/k_{DIS})/4K + N\sigma_{A_2}/2 \qquad\qquad (7a)$$

This enhanced absorption should be manifested in basically two-photon induced fluorescence of monomers photoquenched at higher light intensities. Following standard photoquenching analysis[13] and neglecting contributions from Step 5 (a) we find that the fluorescence intensity I_f is given by

$$I_f = \phi\sigma_{A_2}\sigma_{01}^A AI^2\tau_p/(1 + \tau_{10}\sigma_{01}^A I + \tau_{10}\sigma_{01}^A \tau_{n1}\sigma_{1n}^A I^2) \qquad\qquad (8)$$

where ϕ is the fluorescence quantum yield and τ_p is the laser pulse width.

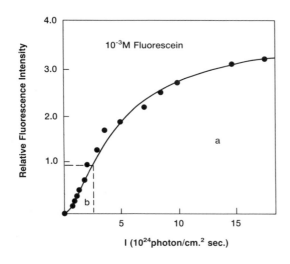

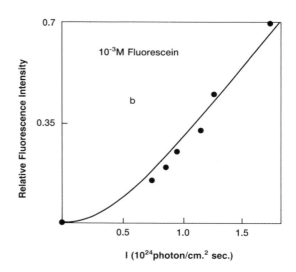

Figure 5a. Laser induced fluorescence intensity for 10^{-3} mole/ℓ fluorescein/ethanol
solution (480 nm excitation). b. A blow up of the $0 - 2 \times 10^{24}$ photon/cm².
sec. region showing the quadratic dependence of the concencutive two-photon
induced fluorescence resulting from dimer photolysis followed by monomer
absorption. The solid line is the best fit to Eq. (8).

   Figure 5 shows a least square fit of Eq. (8) to laser induced fluorescence data for
10^{-3} mole/ℓ fluorescein solutions. The value obtained from the best fit for $\tau_{10}\sigma_{01}$ is in
good agreement with the calculated one. We may thus conclude that at these concentrations
the fluoresceine system is characterized by an absorption coefficient which is an
increasing function of incident laser intensity. At 10^{-6} mole/ℓ, fluorescein is mostly
in the monomer form and laser induced fluorescence studies are typical of photoquenching
process[13] for which

$$I_f = \phi\sigma_{01}^{A}I/(1 + \tau_{10}\sigma_{01}^{A}I + \tau_{10}\tau_{n1}\sigma_{01}^{A}\sigma_{1n}^{A}I^2) \tag{9}$$

This behavior is exemplified in Fig. 6 for fluorescein monomers. The maximum I_f occurs
at[13]

$$I_{max} = (\tau_{10}\sigma_{01}^{A}\tau_{n1}\sigma_{1n}^{A})^{-1/2} \tag{10}$$

The reciprocal relative quantum yield is given by[13]

$$\phi_o/\phi = 1 + \tau_{10}\sigma_{1n}^{A}I \tag{11}$$

   where ϕ_o is the limit of ϕ for $I \rightarrow o$

Figure 7 shows the fit of the experimental relative quantum yield to Eq (11). For $\tau_{10} = 4$ ns
we obtain that $\sigma_{1n}^{A} = 5 \times 10^{-17}$ cm²/molec., and $\tau_{n1} = 2 \times 10^{-10}$ sec.

   Previous calculations have indicated that optical bistability should be observed in
nonlinearly absorbing media.[8,9] This prediction was tested here for the fluorescein
dimer-monomer system. Figure 8(b) shows an increasing absorption optical bistability
hysteresis loop obtained for this system. At low concentrations (10^{-6} mole/ℓ) the loop
disappears, as expected for solutions made of mainly monomer species, resulting in intensity independent α. It is clearly seen that the output pulse peaks before the input pulse
due to increasing $\alpha(I)$. The slower decay of the output pulse (Figure 8(a)) reflects the
recombination rate of monomers, which probably takes place within the solvent cage.

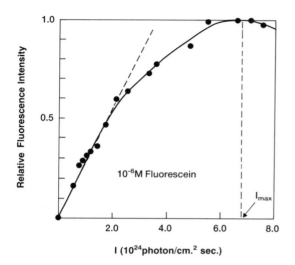

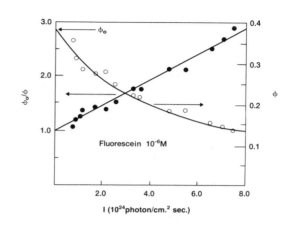

Figure 6. Intensity dependence of laser induced fluorescence of 10^{-6} mole/ℓ fluorescein (480 nm excitation).

Figure 7. Photoquenching plots[13] for 10^{-6} mole/ℓ fluorescein.

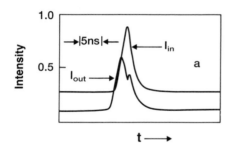

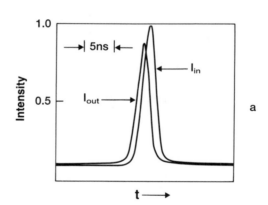

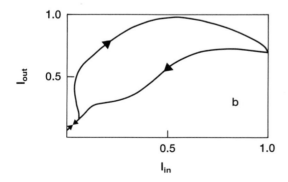

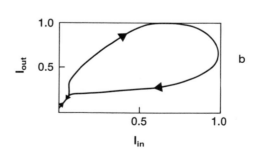

Figure 8. a. Input and output intensity pulse for 10^{-3} mole/ℓ fluorescein in a Fabry-Perot cavity. b. Optical bistability hesteresis loop for the dimer-monomer fluorescein system (480 nm excitation).

Figure 9. Same as Figure 8 for 2 μm thick fluorescein PMMA films (β = 0.64 cm^{-1}).

For fluorescein doped PMMA thin films we observe similar bistable response (Figure 9); however, the decay time of the outpulse is shorter probably due to faster monomer recombination in the more rigid polymer solvent cage.

In conclusion, we note that optical bistability can be observed in monomer-dimer dye systems. Such systems exhibit nonlinear absorption due to photolytic production of monomer species. In this respect it is a kind of photochromic system which was shown to exhibit optical bistability,[14] however on a much longer timescale.

4. REFERENCES

1. J. A. Goldstone, "Optical bistability" in Laser Handbook Vol. 4, M. L. Stitch and and M. Bass eds., North Holland, Amsterdam (1985).
2. H. M. Gibbs, Optical Bistability: Controlling Light With Light, Academic Press, New York (1985).
3. A. Szoke, V. Dannell, J. Goldhar, and N. A. Kurnit, "Bistable optical element and its applications" Appl. Phys. Lett. 15, 376-379 (1969).
4. H. Seidel, U.S. Patent No. 3610731 (1969).
5. S. L. McCall, "Instabilities in continuous-wave light propagation in absorbing media" Phys. Rev. A9, 1515-1518 (1974).
6. M. Orenstein, S. Speiser and J. Katriel, "An eikonal approximation for nonlinear resonators exhibiting bistability" Opt. Commun. 48, 367-373 (1984).
7. M. Orenstein, S. Speiser and J. Katriel, "A general eikonal treatment of coupled dispersively nonlinear resonators exhibiting optical multistability" IEEE J. of Quant. Elec. QE-21, 1513-1522 (1985).
8. M. Orenstein, J. Katriel and S. Speiser, "Nonlinear complex eikonal approximation: optical bistability in absorbing media" Phys. Rev. A 35, 1192-1209 (1987).
9. M. Orenstein, J. Katriel and S. Speiser, "Optical bistability in molecular systems exhibiting nonlinear absorption" Phys. Rev. A 35, 2175-2183 (1987).
10. R. W. Chambers, T. Kajiwara and D. R. Kearns, "Effect of dimer formation on the electronic absorption and emission spectra of ionic dyes. Rhodamine and other common dyes" J. Phys. Chem. 78, 380-387 (1974).
11. I. Lopez Arbeloa, "Dimeric and trimeric states of the fluorescein dianion" part 1, J. Chem. Soc., Faraday Trans. 2. 77, 1725-1733; part 2, ibid. 1735-1742 (1981).
12. W. E. Ford, "Photochemistry of 3,4,9,10-perylenetetracarboxylic dianhydre dyes: absorption and fluorescence of the di(glycyl)imide derivative monomer and dimer in basic aqueous solutions" J. Photochem. 37, 189-204 (1987).
13. S. Speiser and N. Shakkour, "Photoquenching parameters for commonly used laser dyes" Appl. Phys. B38, 191-197 (1985) and references therein.
14. C. J. G. Kirkby, R. Cush and I. Bennion, "Optical nonlinearity and bistability in organic photochromic thin films" Opt. Commun. 56, 288-292 (1985).

ADVANCES IN NONLINEAR POLYMERS AND INORGANIC CRYSTALS,
LIQUID CRYSTALS, AND LASER MEDIA

Volume 824

Session 4

Devices and Applications of the New Materials

Chair
R. Lytel
Lockheed Missiles & Space Company, Inc.

Invited Paper

Advances in organic electro-optic devices

R. Lytel, G.F. Lipscomb, J. Thackara, J. Altman, P. Elizondo, M. Stiller, and B. Sullivan

Lockheed Research & Development Division, Lockheed Missiles & Space Company, Inc.
3251 Hanover St., 0/97-20, B/202, Palo Alto, California 94304

ABSTRACT

Organic and polymeric materials have many useful features for thin-film electro-optic devices. These include low dielectric constants, moderate-to-large electro-optic coefficients, and low optical loss. This paper presents a review of the useful features of organic materials for device applications, summarizes the current optical response levels of some organic materials, and describes the performance of several device prototypes fabricated from poled organic films.

1. INTRODUCTION

Organic and polymeric materials have emerged in recent years as promising candidates for advanced device and system applications. This interest has arisen from the promise of extraordinary optical, structural, and mechanical properties of certain organic materials, and from the fundamental success of molecular design performed to create new kinds of materials.[1] From an optical standpoint, organics offer temporal responses ranging over 15 orders of magnitude, including large nonresonant electronic nonlinearities (fs-ps), thermal and motional nonlinearities (ns-μs), configurational and orientational nonlinearities (μs-s), and photochemical nonlinearities (ps-s). Additionally, organic and polymeric materials can exhibit high optical damage thresholds, broad transparency ranges, and can be polished or formed to high-optical-quality surfaces. Structurally, materials can be made as thin or thick films, bulk crystals, or liquid and solid solutions, and can be formed into layered film structures, with molecular engineering providing different optical properties from layer to layer. Mechanically, the materials can be strong and resistant to radiation, shock, and heat. When coupled with low refractive indices and dc dielectric constants, the collective properties of these extraordinary materials show great promise toward improving the performance of existing electro-optic and nonlinear optical devices, as well as allowing new kinds of device architectures to be envisioned.

However, as with any new class of materials, the existence of certain promising samples does not imply that real applications will necessarily be possible. Real optical materials must exhibit some basic properties, including optical clarity (very low scattering and absorption losses), fabricability, and the potential for mass production. These secondary properties are generally not addressed by fundamental research, but it is the secondary properties which will determine whether the materials can have any practical use. For organics as a class, including pi-electron systems, conducting polymers, and other nonlinear optical polymers, a great deal of research and development remains to achieve usable, exciting new materials.

This paper provides a survey of the current research and development underway at Lockheed in organic and polymeric devices. In particular, we examine organics as a new class of nonlinear and electro-optic (E-O) materials, delineate their good and bad features, and present a discussion of certain existing organic materials, their use in devices, and their impact on device performance. Along the way, we try to provide the motivation for using an organic material rather than some other material, and point toward the specific new fabrication requirements for devices, and toward the further materials development required for optimum device performance. Our approach is a positive one, as we are firm believers that organics will have a major impact on the optoelectronic device technology of the next decade. However, we moderate our optimism with the realization that much materials research and development remains to be done before organics can even be established as an important new class of practical nonlinear optical materials.

Section 2 reviews some of the important structural and optical properties of organic materials for nonlinear optical devices, and points toward specific properties that have been demonstrated, and those that remain to be proved. Section 3 details the application of certain organic and polymeric materials to practical optical devices, including photo-addressed spatial light modulators, integrated optical devices, and gratings. We review our own work in these areas, and point toward specific performance and manufacturing advantages of organics over inogranic materials. Section 4 summarizes our conclusions regarding organic and polymeric materials for device applications, and provides some directions for future research in this exciting new area of nonlinear optical devices and applications.

2. REVIEW OF MATERIAL PROPERTIES

For the purposes of this paper, we shall refer by organics to nonresonant pi-electron organic and polymeric materials, including single crystals, thin-films, and other composite materials. As a class, organics offer a number of exciting optical structural properties for devices, summarized in Table 1. Some of these properties, such as the capability to alter linear and nonlinear optical properties in dimensions approaching visible optical wavelengths, are the result of the ability to molecularly engineer the active NLO units within a given structure.[2] This feature appears to be unique to organics, although molecularly tailored semiconductor multiple quantum wells offer some tunability in wavelength and response. However, the capability to engineer materials with desired linear and nonlinear properties is unique to organics, and is the major reason they are, as a class, important for device applications.

Table 1. Some Structural and Optical Properties of Organics

STRUCTURAL	OPTICAL
• MOLECULAR ENGINEERING	• LARGE NONRESONANT NONLINEARITIES
• THIN FILMS/BULK XTAL	• LOW dc DIELECTRIC CONST.
• ROOM TEMPERATURE	• FAST NLO RESPONSE
• CHEMICAL/STRUCTURAL STABILITY	• HIGH OPTICAL DAMAGE THRESHOLDS
• INTERNAL GRATINGS	• BROADBAND
• INTEGRATED OPTICS	• LOW ABSORPTION

Many of the features in Table 1 have been demonstrated in a number of different organic systems. Damage thresholds as high as a GW/cm^2 have been measured in single crystal MNA[3] (2-methyl-4-nitroaniline) and in urea.[4] Large nonresonant susceptibilities have been reported in certain diacetylene polymer systems[5] ($\chi^{(3)}$ of the order of 10^{-10} to 10^{-9} esu) and in single crystal MNA[6] ($\chi^{(2)}$ = 500+/-100 pm/V). Other recent measurements of large electro-optic coefficients have been reported in poled organic films,[7] which show great promise for optical waveguides. It is noteworthy that current synthesis efforts underway are moving away from crystal growth and toward thin-film fabrication of nonlinear polymers, and that none of these systems has a $\chi^{(3)}$ or $\chi^{(2)}$ as large as the original reported susceptibilities of the diacetylenes and MNA crystals. For all of the problems of growing crystals, they yield the greatest degree of orientation, and, therefore, the largest macroscopic nonlinearities for a given NLO moiety.

It is also true that most organics fabricated to date do not yet exhibit most or all of the properties in Table 1. In particular, absorption and scattering losses are usually higher than desired, and must be minimized by building materials in cleanroom facilities or by filtering polymer solutions before using them. Many polymer films exhibit poor optical quality, and techniques for fabrication of optically flat surfaces need to be refined. Some polymer structures are inherently grainy or contain fibrous components which scatter light excessively. Finally, a few researchers are reporting new materials, such as conducting polymers, with large $\chi^{(3)}$ values, without recognizing the requirement that the material must inherently be able to transmit or guide light! These reports indicate a lack of understanding of what constitutes an optical material, as opposed to an electrical material. Nonetheless, we remain confident that the field of organics will develop into an important new area of device applications, once the materials and device scientists jointly address the problem.

In this light, it is straightforward to examine the current status of $\chi^{(2)}$ and $\chi^{(3)}$ materials,[8] and to determine what materials parameters ought to be improved for real applications.

2.1. SHG and electro-optic materials

The largest reported SHG[9] and E-O[6] coefficients in an organic material exist in the organic solid MNA, in single crystal form. This result is interesting for several reasons. First, while the NLO moiety, the MNA molecule, has a large second-order molecular hyperpolarizability β, many other molecules with larger β have been reported. However, single crystals of MNA have been grown and characterized, illustrating the effect of macroscopic orientation on the macroscopic polarizability [2]. Thus, we learn that it is critical that proper orientation be achieved to take advantage of the large β in certain molecular systems. This just illustrates the difference between a molecule and a material. Recent results in poled polymer films suggest that we are well on our way toward achieving large, usable $\chi^{(2)}$'s from partially oriented films of different polymer systems. Both AT&T and Celanese[7] have reported new polymers exhibiting interesting $\chi^{(2)}$'s, and it is likely that materials currently unreported upon exist and have even greater $\chi^{(2)}$'s. In our view, it is only a matter of time before poled polymer films will appear regularly in real optical devices. As reported below, we have already used certain poled films to produce "proof of concept" prototypes of optical waveguide modulators with great success.

A major drawback to the current poled polymer films is that the $\chi^{(2)}$ is perpendicular to the film surface. For waveguides, this is no problem. However, for bulk E-O device applications, it is desirable to have a film with $\chi^{(2)}$ parallel to the film surface to achieve E-O modulation for two-dimensional device applications. It remains to be seen whether new orientation techniques, equivalent to surface alignment techniques for nematic liquid crystals, can be developed.

There is one major observation worthy of note when comparing organic pi-electron materials with inorganic crystals in terms of their $\chi^{(2)}$. In both materials, the contributions from $\chi^{(2)}$ to SHG arise solely from the electronic contributions, since these are the only ones capable of responding on femtosecond time scales. In organics, the same contributions also contribute to the entire E-O coefficient, while inorganics gain much of their contribution from lattice phonons. Thus, an organic material with the same $\chi^{(2)}$ as an inorganic material, as measured by SHG, will probably have a much smaller E-O coefficient than the inorganic material. This result must be borne in mind when comparing organics, as characterized by SHG, with E-O inorganic materials.

2.2. Third-order materials

The largest nonresonant $\chi^{(3)}$ yet reported[5] for an organic material is in the diacetylene systems, and has a $\chi^{(3)}$ of the order of 10^{-9} esu. This value is large by comparison with CS_2, but is still too small for most applications. In devices based on $\chi^{(3)}$, it is generally the product of $\chi^{(3)}$ with the light-medium interaction length and the optical intensity that determines the net nonlinear phase shift that can be obtained and exploited for device operation. Small $\chi^{(3)}$ necessarily implies large optical intensities and/or long interaction lengths. For nonresonant organics like PTS diacetylene polymers, intensities of order MW/cm^2 would be required over a length L=1 cm. Thus, it is likely that third-order organics will find applications in optical waveguide devices, such as bistable optical switches and optically controlled modulators and couplers. It is unlikely that thin-film applications, such as etalons, will be possible with organics unless significant advances toward achieving larger nonlinearities can be made.

Organic thin films can, however, offer some really new features for third-order devices, if the coefficients can be made larger. Through molecular engineering, it should be possible to produce anisotropic $\chi^{(3)}$ materials for optically activated birefringent film applications. It should even be possible to build layered structures with a $\chi^{(3)}$ that varies along the thickness of the layers, thus producing nonlinearly activated optical gratings for fixed and tunable filter applications.

Recent synthesis work[10] includes off-resonance side and main chain polymers with reasonable nonlinearities (10^{-11} to 10^{-10} esu). Such materials could be useful in degenerate four-wave mixing[11] for optical phase conjugation, or as self-focusing media for optical shutters. However, it is still necessary to achieve optically good materials (clarity and optical flatness) and larger susceptibilities for most applications.

In summary, second-order poled polymer films appear to be close to achieving interesting levels of response for waveguide E-O device applications. However, third-order NLO polymers still have a way to go. In both cases, the materials must necessarliy qualify as optical materials before real applications can be envisioned. In light of the recent success in producing poled polymer films, we discuss next three important device application areas for these materials.

3. DEVICE APPLICATIONS

In this section, we report recent work toward achieving useful devices based on second-order, poled polymer films. The poling process produces a material with a $\chi^{(2)}$ perpendicular to the film surface. Our major experimental results to date have been obtained in E-O waveguide devices using a polymer known to us as PC6S, and provided by the

Hoechst Celanese Research Company. This polymer is optically transparent above 0.6 μm, and has a measured value of r_{33} = 2.8 pm/V. We report next on our investigations into specific devices for exploiting poled polymer films.

3.1. Spatial light modulators

Spatial light modulators (SLMs) are two-dimensional optical modulation devices.[12] Such devices are of great interest in optical processing, computing, and beam control, and are usually photo-addressed devices. A typical electro-optic SLM, operating in reflection, is illustrated in Fig. 1. This device is an E-O modulator, composed of a photoreceptor, dielectric mirror, and an E-O material. The SLM operates as follows: An incident optical field addresses the photoreceptor, creating a two-dimensional charge distribution which is proportional to the intensity of incident light. Under the influence of a bias voltage, the charge distribution $\sigma(x,y)$ migrates to the dielectric mirror-photoreceptor interface, and modulation of the E-O material due to the charge distribution can occur. If the modulation is due to the difference between σ (and the bias voltage) and the ground plane, the device is a longitudinal modulator. If the modulation is due to local differences in σ at the surface of the E-O material, the device is a transverse modulator. In the longitudinal modulator, a readout field incident from the right acquires a phase shift proportional to the induced E-O modulation due to $\sigma(x,y)$, and the device operates as an intensity to phase converter. In the transverse device, it is the local field gradient along the dielectric mirror and E-O material interface which creates the modulation field, and the device produces a phase shift upon readout proportional to the gradient of the intensity. Thus, the fundamental operation of the E-O modulator depends on the tensorial nature of the E-O medium in a fundamental way.

Electro-optic SLMs can be seen to be important optical devices because they allow the transfer of optical information from one beam to another. In particular, a longitudinal modulator can be used as an optical correlator, convolutor, or optical phase conjugator, while a transverse modulator can be used as an edge detector or intensity-to-position encoder. In either case, SLM performance parameters are determined primarily by the optical properties of the photoreceptor and E-O material, and the electrical properties of the entire unit.

Photo-addressed E-O SLM technology has been under development at Lockheed for the past several years. The Lockheed photoreceptor is a high-resistivity silicon device, with very fast response times (of the order of a ns) and good lateral charge confinement. Lockheed has mated this technology to thin (100-μm) KD*P crystals to produce fast E-O modulation devices with spatial resolution approaching 10 line pairs/mm, frame grabbing times under 10 μs, and frame rates approaching a kHz.[13] This architecture has formed the basis for all of our subsequent SLM prototypes, including a longitudinal nematic liquid crystal device,[14] a transverse nematic liquid crystal edge detection device,[15] and the current organic transverse SLM under development.

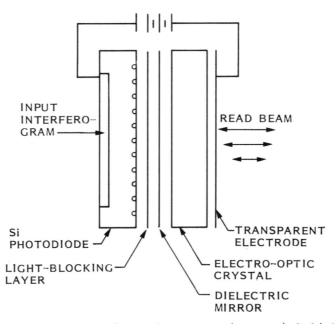

Fig. 1. Schematic diagram of an electro-optic spatial light modulator

The impact on using an organic material in the Lockheed SLM architecture is currently under experimental evaluation, and no results are available at the time of this paper. However, we have evaluated theoretical performance for an organic SLM, and it is worthwhile to explore this briefly to understand the importance of organics in SLM technology.

In the Lockheed device, response time for a single frame is determined by the integration time of the input light required to generate sufficient charge to get a desired modulation (say half-wave) on the E-O material, the RC time constant of the device, and the intrinsic response time of the E-O material. The response of the KD*P is subnanosecond, so the device frame time is determined solely by the write light intensity and the device capacitance. The KD*P is operated near its Curie temperature of -57°C, so that the enhancement in the E-O coefficient can be obtained, allowing the bias voltage to be kept below the breakdown voltage of the photoreceptor. The dielectric constant of KD*P, however, also rises dramatically near the Curie temperature and no benefit in device efficiency is achieved. Since the polarization optic coefficient is not strongly temperature dependent, the same amount of photo-generated charge (σ) is required to produce an equivalent phase modulation, and the required integration time is not reduced. Large dielectric constants are an inescapable feature of inorganic ferroelectric E-O materials, arising from the electron-phonon coupling origin of the electro-optic effect. The resultant large device capacitance adversely affects the speed and sensitivity that can be achieved in a photo-addressed device. Finally, device resolution is limited, by capacitative effects, to about 10 lp/mm. Organic thin films exhibiting a large E-O coefficient and a small dielectric constant should offer significant improvements.

A nematic liquid crystal version of the same device has been developed at Lockheed, in the hopes of achieving much more sensitiviy and greater spatial resolution. This device, operated at room temperature, exploits the birefringence induced by an applied electric field. The nature of the device depends on the surface alignment of the liquid crystal. Parallel alignment produces a longitudinal modulator, while perpendicular alignment produces a transverse modulator. In both cases, a significant gain in device sensitivity is achieved, due to the much lower dielectric constant of the nematic liquid crystal relative to cooled KD*P and the larger birefringence of the nematic liquid crystal. Higher resolution is achievable because thinner films can be utilized. However, the nematic alignment must be driven off to achieve millisecond response times, and this device is never as fast as the KD*P device.

Organic E-O films can probably achieve higher resolution, higher speed, and greater sensitivity in an E-O SLM than either inorganic E-O crystals or liquid crystals. A poled organic film of the order of 10-μm thick, with an E-O coefficient comparable to KD*P, and a dielectric constant $\epsilon=4$, should produce a longitudinal SLM with sensitivity comparable to a nematic liquid crystal device, speed of a KD*P device, and resolution much better than both KD*P and nematic liquid crystals. This increased performance results from the lower film capacitance at room temperature for a given E-O coefficient. Such films, available now as poled polymer films oriented perpendicular to the surface of the film, cannot be used in normal incidence and must be read out at an angle. We expect to have the first prototype SLM of this type operating by the end of 1987.

A major drawback to the use of organics in E-O devices will arise from the poor thermal conductivity of polymers. Thus, repetition rates and optical loads will be limited by the device architecture, and careful design of heatsinks and thermal conditions will be necessary. If properly designed, the ultimate optical load of the device should be much greater than inorganic devices, due to the potentially larger optical damage threshold of organic materials.

3.2. Integrated optical devices

Organic E-O materials offer a variety of potential advantages over conventional materials for integrated optical device applications. Table 2 provides a comparison of the potential of organic materials with the current frontrunner technology, Ti-indiffused LiNbO$_3$ in three major areas of importance: materials parameters, processing technology, and fabrication technology. Some of these advantages have already been realized in our work on E-O modulators using poled polymer films.

The most obvious potential advantages are due to the intrinsic differences in E-O mechanisms in inorganic and organic materials. Organics should provide flat E-O response well beyond a GHz, and, indeed, measurements of the E-O coefficients and SHG coefficients of certain poled polymer films show little or no dispersion. Second, it is likely that the E-O coefficients of poled polymer films can be made nearly as large as LiNbO$_3$, and it is obvious that many materials efforts underway in the U.S. and abroad are attempting to achieve exactly that. Pure MNA crystals already exhibit larger E-O coefficients than LiNbO$_3$, as indicated in Table 2, but it is our sense that real device advances will be made with films, not crystals. It is also true that the dielectric constant of poled poly-

Table 2. Comparison of Integrated Optics Technologies:
Current Ti-LiNbO$_3$ and Projected Organics Technologies

- CURRENT TECHNOLOGY: Ti:LiNbO$_3$
    - MATERIALS DEVELOPMENT BEGAN IN 1930s

- r = 32 pm/V
    - LARGER MODULATING VOLTAGE
    - LITTLE IMPROVEMENT EXPECTED

- LIMITED FABRICABILITY
    - 1000°C PROCESSING
    - DEPTH LIMITED TO 5 μm
    - LOW INDEX CHANGE Δn
    - LOSS > 0.1 dB/cm
    - OPTICAL DAMAGE (PHOTOREFRACTOR)

- LARGE DIELECTRIC CONSTANT (28)
    - LONGER TIME CONSTANTS = RC
    - LARGE VELOCITY MISMATCH IN TRAVELING WAVE MODULATOR

- MASS PRODUCTION DIFFICULT

- POTENTIAL ADVANTAGES OF ORGANIC E-O MATERIALS
    - MATERIALS DEVELOPMENT BEGAN IN 1975

- r = 67 $\pm$ 25 pm/V (MNA)
    - LOWER MODULATING VOLTAGE
    - POTENTIALLY MUCH LARGER r

- FLEXIBLE FABRICATION
    - LOW TEMPERATURE PROCESSING
    - FLEXIBLE DIMENSIONS
    - CONTROLLABLE INDEX CHANGE Δn
    - POTENTIAL LOW LOSS
    - HIGH OPTICAL DAMAGE THRESHOLD

- LOW DIELECTRIC CONSTANT (4)
    - SHORTER TIME CONSTANTS = RC
    - SMALLER VELOCITY MISMATCH

- POTENTIAL FOR MASS PRODUCTION

mer films is substantially lower than that of LiNbO$_3$, implying smaller RC time constants and wider frequency bandwidths. Finally, the processing technology for integrated optical devices based on poled polymer films is relatively straightforward and fast, requiring only moderate temperatures (100°C) for poling, and standard semiconductor fabrication equipment for fabrication of layered waveguide structures.

As with all other materials, the secondary issues will be the drivers for organic integrated optical devices. Low absorption and scattering losses need to be designed into the material and fabrication technology, and good quality films must be regularly producible. However, it seems likely that these achievements can be met with further research and development. It is worth noting that the current inorganic crystal growth technology is a very old technology, representing thousands of man-years of research. Much work remains to be done with organics.

The Lockheed organic integrated optical devices effort has recently focused on the fabrication of simple slab guided wave structures made from polymer films. These structures allow the determination of fabrication techniques for organic devices, as well as direct measurement of Kerr and E-O coefficients in the waveguide configuration. By way of example, we next discuss two guided wave structures made from organics: a Kerr effect modulator based on an unoriented MNA film and an electro-optic modulator based on the PC6S poled polymer film described earlier.

3.2.1. Kerr-effect modulator. The first organic guided wave device fabricated at Lockheed was a Kerr effect modulator based on an MNA/PMMA guest-host film.[16] This structure, illustrated in Fig. 2, is a slab modulator consisting of a thin film of SiO$_2$ as a lower buffer layer, the MNA/PMMA film as the guide, and another thin buffer layer film composed of polysiloxane. The entire structure is built on an aluminum-coated glass substrate with an aluminum electrode at the top. Layer dimensions and composition are illustrated in Fig. 3. An HeNe laser was prism-coupled into the device, which guided both a TE and TM mode. Intermode interference was observed on output, and the frequency response of the device is illustrated in Fig. 4. The low-frequency response is exactly what one would expect from a Kerr effect in this material. The output was measured with a lock-in amplifier, set to lock to twice the modulation frequency of the applied voltage, as appropriate for a Kerr effect. A linear effect was also sought, but was not measurable, as expected for an unpoled film.

Although this device is not useful for applications, its fabrication helped define procedures for the fabrication of E-O slab waveguide devices, described next.

3.2.2. Electro-optic modulator. The next slab modulator constructed at Lockheed consisted of a layered structure similar to that in Fig. 3, with the PC6S polymer film replacing the MNA/PMMA film.[16] The poling procedure consisted of first spin-coating the electrode-coated glass with the bottom buffer layer and then the PC6S, applying a top electrode, and applying a voltage of the order of 1 MV/cm to the structure. Poling was observed with a

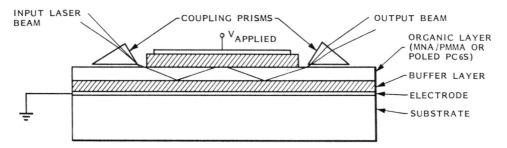

Fig. 2. Schematic of Kerr-effect waveguide modulator

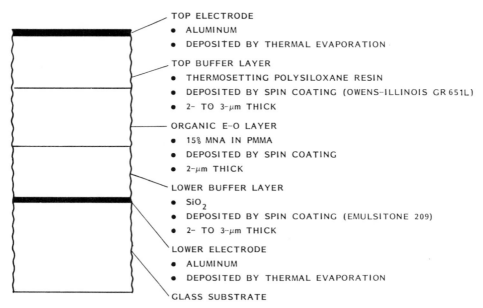

TOP ELECTRODE
- ALUMINUM
- DEPOSITED BY THERMAL EVAPORATION

TOP BUFFER LAYER
- THERMOSETTING POLYSILOXANE RESIN
- DEPOSITED BY SPIN COATING (OWENS-ILLINOIS GR 651L)
- 2- TO 3-μm THICK

ORGANIC E-O LAYER
- 15% MNA IN PMMA
- DEPOSITED BY SPIN COATING
- 2-μm THICK

LOWER BUFFER LAYER
- SiO_2
- DEPOSITED BY SPIN COATING (EMULSITONE 209)
- 2- TO 3-μm THICK

LOWER ELECTRODE
- ALUMINUM
- DEPOSITED BY THERMAL EVAPORATION

GLASS SUBSTRATE

Fig. 3. Layers composing Kerr-effect modulator

polarizing microscope, and the entire procedure was monitored in realtime to optimize the poling. The top electrode was then removed, and a top buffer layer and new electrode were applied to the poled structure. Guiding of 830-nm light from a semiconductor laser over a 1.5-cm dimension with minimal loss was then achieved by locating the prism couplers directly over the ends of the poled region.

The optical response was measured interferometrically, and is illustrated in Fig. 5. The device half-wave voltage was about 23 V. Figure 6 illustrates the measured frequency response out to about 100 kHz, and shows it to be flat. The E-O coefficient of the poled PC6S film was measured to be 2.8 pm/V. Higher frequency measurements are currently underway with a new optical test system, and will be reported later in 1987.

The fabrication of such poled slab waveguide devices from spin-coated polymer films represents the first major steps toward the development of organic integrated optical devices. Many issues remain to be studied such as measurement and minimization of scattering losses, measurement of higher frequency response, development of procedures to form two-dimensional guides, and development of simple optoelectronic devices. However, the implications of this work are significant and promising. Simple fabrication techniques, requiring only standard spin-coating tools and chemical etching, can produce organic electro-optic devices from unoriented polymer films.

3.3. Fabrication of gratings for devices

Organic and polymeric materials are ideal for the fabrication of gratings as component parts of more complex electro-optic and nonlinear optical devices. Optical gratings have been fabricated in polymers by partially polymerizing monomers in crossed UV beams.[17] Gratings of variable spacing can be fabricated in this manner, providing couplers for waveguide devices, Bragg mirrors for high-finesse etalon devices,[11] and holograms for optical memory and data storage. The unique processing of polymer structures, viz., spin-coating of films and subsequent poling of active E-O layers, can lead to the development

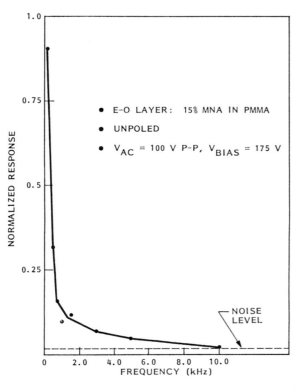

Fig. 4   Measured frequency response of
Kerr-effect modulator

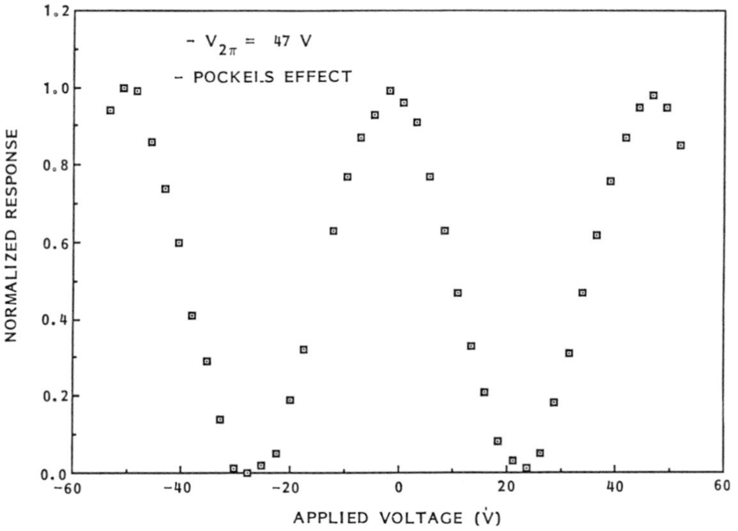

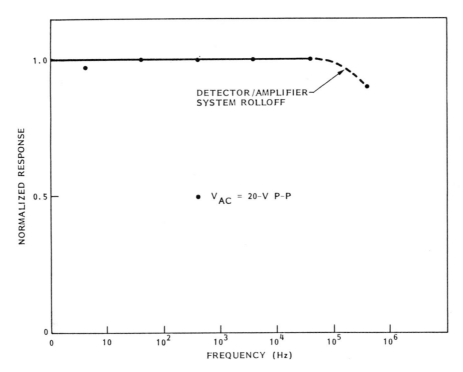

Fig. 5.   Interferometric measurement of E-O
modulator response

Fig. 6.   Frequency response of E-O waveguide modulator

of true, integrated optic devices with couplers, guide layers, and active E-O regions
integrated into a single thin-film sandwich.

We have begun preliminary work on the development of gratings for integrated optic
devices.   Rather than polymerizing monomer films, we have adapted a photo-active Polaroid

film, DMP-128, for use in the fabrication of gratings for Bragg mirrors and waveguide couplers. Figure 7 illustrates the fabrication procedure. The film is illuminated by a UV interferogram to provide a spatially varying, sinusoidal intensity distribution in the material. After exposure, the film is processed and fixed. The resultant structure contains a sinusoidally varying refractive index $n(z) = n_0 + n_1 \sin(2\beta_0 z)$, due to the chemical process involved in the exposure. Here, β_0 is the grating wave vector, and z is the direction normal to the film. A small refractive index modulation depth n_1/n_0 of only a few percent in a 25-μm-thick film is sufficient to produce a Bragg mirror with a reflectivity of over 90 percent for normally incident light.[18] This type of processing has also been used to develop waveguide grating couplers for the devices described previously in this paper. Future research will address the fabrication of etalons using third-order, resonant films with Bragg mirrors fabricated in this manner.

- INTENSITY DEPENDENT PHOTOCHEMICAL REACTIONS RESULT IN LOCAL CHANGES IN THE INDEX OF REFRACTION OF THE MEDIUM

- ALLOWS THE ENCODING OF OPTICAL PHASE INFORMATION

- EXAMPLE: SIMPLE HOLOGRAPHIC GRATING FORMATION

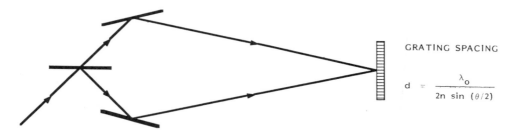

GRATING SPACING

$$d = \frac{\lambda_0}{2n \sin(\theta/2)}$$

Fig. 7. Fabrication of gratings in photo-active polymer films

4. CONCLUSIONS

We have described certain features of organic and polymeric materials and their applications to nonlinear and electro-optic devices. As a class, the materials offer an important number of structural and optical properties for device applications, including flexible and straightforward fabricatibility, large E-O coefficients, low dc dielectric constants, and high optical damage thresholds. For second-order materials, current advances indicate that materials with performance approaching that of $LiNbO_3$ should soon be available, and a major impact on E-O device technology will be forthcoming. For third-order materials, it is likely that guided wave devices, with their long interaction lengths, will be the most practical for organics. At this time, it is still true that MNA (second order) and diacetylene polymers (third order) have the largest optical nonlinearities, as it was when original results were reported. However, it is clear that the fundamental material science is producing a working knowledge of how to engineer materials with larger nonlinearities, and it is only a matter of time before organics emerge as a promising class of practical nonlinear optical materials.

It is our view that future research in $\chi^{(3)}$ materials be carefully examined to ensure that claims of large nonlinearities in optically poor media do not spoil the real progress being made in the development of new materials. Nonlinearities of the order of 10^{-9} esu are not large at all. Instead, they are large for a nonresonant material, but still imply devices with large operating intensities. Nonresonant nonlinearities offer broadband response, but operating intensities must be reasonable, and should be obtainable from diode laser sources, as the latter can already be used with the large resonant response of III-V semiconductor multiple quantum well structures (10^{-1} esu). Care must be taken not to overpromote the $\chi^{(3)}$ of a new material without first carefully comparing its material parameters to other materials, and then comparing its performance in a device. As we have seen, device performance depends on materials parameters in a very complicated way, and is often determined by the device architecture or electrical properties, rather than the material.

5. ACKNOWLEDGMENTS

Our research is a cooperative effort with the Hoechst-Celanese Research Company, and we wish to acknowledge their excellent cooperation, support, and teamwork. In particular, we thank Dr. James Stamatoff and his group for providing samples, for working with us to develop new device fabrication techniques, and for his equal enthusiasm for the research underway in this new field.

6. REFERENCES

1. For a recent review, see Nonlinear Optical Properties of Organic Molecules and Crystals, Vols. 1 and 2, D. Chemla and J. Zyss, ed., Academic Press, MO (1986). See also S.J. Lalama and A.F. Garito, "Origin of the nonlinear second-order optical susceptibilities of organic systems," Phys. Rev. A 20, 1179 (1979); and "Nonlinear optical properties of organic and polymeric materials," D.J. Williams ed., ACS Symposium Series 233, American Chemical Society, (1983).

2. A.F. Garito, "Nonlinear optics: organic and polymeric systems," Proc. American Chemical Society Symposium on Electro-Active Polymers, Denver, CO, April, 1987 (to be published).

3. A.F. Garito, private communication.

4. C. Cassidy, J.M. Halbout, W. Donaldson, and C.L. Tang, "Nonlinear optical properties of urea," Opt. Comm. 29, 243 (1979).

5. C. Sauteret, J.-P. Hermann, R. Frey, F. Pradere, and J. Ducuing, "Optical nonlinearities in one-dimensional conjugated polymer crystals," Phys. Rev. Lett. 36, 956 (1976); G.M. Carter, Y.J. Chen, and S.K. Tripathy, "Intensity-dependent index of refraction in multilayers of polydiacetylene," Appl. Phys. Lett. 43, 891 (1983).

6. G.F. Lipscomb, A.F. Garito, and R.S. Narang, "An exceptionally large linear electro-optic effect in the organic solid MNA," J. Chem. Phys. 75, 1509 (1981).

7. C.S. Willand, S.E. Feth, M. Scozzafava, D.J. Williams, G.D. Green, J.I. Weinshenk, H.K. Hall, and J.E. Mulvaney, "Electric-field poling of nonlinear optical polymers," J.B. Stamatoff, A. Buckley, G. Calundann, E.W. Choe, R. DeMartino, G. Khanarian, T. Leslie, G. Nelson, D. Stuetz, C.C. Teng, and H.N. Yoon, "Development of polymeric nonlinear optical materials," K.D. Singer, J.E. Sohn, and M.G. Kuzyk, "Orientationally ordered electro-optic materials," Proc. American Chemical Society Symposium on Electro-Active Polymers, Denver, CO, April, 1987 (to be published).

8. It is not as straightforward as we imply to evaluate the status of nonresonant organic materials. We feel certain that the latest industrial developments are unreported, especially for second-order materials, because they should have significant value to the developers.

9. B.F. Levine, C.G. Bethea, C.D. Thurmond, R.T. Lynch, and J.L. Bernstein, "An organic crystal with an exceptionally large optical second-harmonic coefficient: 2-methyl-4-nitroaniline," J. Appl. Phys. 50, 2523 (1979).

10. A.C. Griffin, "Side-chain liquid crystalline copolymers for NLO response," Proc. American Chemical Society Symposium on Photo-Active Polymers, Denver, CO, April 1987 (to be published).

11. G.F. Lipscomb, J Thackara, R. Lytel, J. Altman, P Elizondo, E. Okasaki, M. Stiller, and B. Sullivan, "Optical nonlinearities in organic materials: fundamentals and device applications," Proc. SPIE Vol. 682, 125 (1986).

12. For a comprehensive review, see Optical Engineering, Vol. 25, No. 2 (1986).

13. D. Armitage, W.W. Anderson, and T.J. Karr, "High-speed spatial light modulator," IEEE J. Quant. Elec. QE-21, 1241 (1985).

14. D. Armitage, J.I. Thackara, W. Eades, M. Stiller, and W.W. Anderson, "Fast nematic liquid-crystal spatial light modulator," submitted to SPIE San Diego, August, 1987.

15. D. Armitage and J.I. Thackara, "Liquid-crystal diffrentiating spatial light modulator," Proc. SPIE 613, 165 (1986); "Ferroelectric liquid-crystal and fast nematic spatial light modulators," D. Armitage, J.I. Thackara, N.A. Clark, and M.A. Handschy, Proc. SPIE 684, 60 (1986).

16. J.I. Thackara, G.F. Lipscomb, R. Lytel, M. Stiller, E. Okasaki, R. DeMartino, and H. Yoon, "Optoelectronic waveguide devices in thin-film organic media," CLEO-87, paper ThK29 (1987).

17. K.H. Richter, W. Guttler, and M. Schierer, "UV-holographic gratings in TS-diacetylene single crystals," Appl. Phys. A32, 1 (1983).

18. R. Lytel and G.F. Lipscomb, "Narrow-band electro-optic tunable notch filter," Appl. Opt. 25, 3889 (1986).

OPTICAL POWER LIMITING BEHAVIOR OF
NOVEL NONLINEAR OPTICAL POLYMERS

Saukwan Lo[1], Samson A. Jenekhe[2], and Stephen T. Wellinghoff [3]

1   Honeywell Inc., Systems and Research Center
    3660 Technology Drive
    Minneapolis, Minnesota 55418

2   Honeywell Inc., Physical Sciences Center
    10701 Lyndale Avenue South
    Bloomington, Minnesota 55420

3   Southwest Research Institute
    Division of Chemistry & Chemical Engineering
    6220 Culebra Road
    San Antonio, Texas 78284

ABSTRACT

Optical power limiting behavior is demonstrated in several nonlinear optical polymers and organic model compounds that exhibit third-order optical nonlinearities. A remarkably sharp intensity-limiting behavior was observed in nonlinear transmission measurements of output versus input intensities at 1064 nm laser light without an aperture. The soluble novel organic polymers were found to exhibit both absorptive and refractive nonlinear optical effects that were evidenced by nonlinear absorption and temporal pulse steepening measurements. Nonlinear absorption coefficient (α_2) and nonlinear index of refraction (n_2) as high as 10^{-7} cm/W and 10^{-11} cm^2/W respectively were obtained at 1064 nm using a nanosecond Q-switched Nd:YAG laser. The observed nonlinearities decreased when measured with a 35 picosecond mode-lock laser. However, since the observed nonlinearities were independent of the energy density of the laser irradiation, it is suggested that the nonlinear mechanism is not thermal in origin. Solutions and solid films of the organic materials were studied. Intensity-limiting thresholds as low as 10 MW/cm^2 and laser damage threshold greater than 10 GW/cm^2 were obtained. It is suggested that nonlinear optical polymers hold promise for applications in passive optical limiting and regulating devices.

I. INTRODUCTION

Optical power limiters that are frequency independent are clearly useful as a protection method for sensitive optical components over a broad spectral range. Among the important advantages of such a protection method over fixed wavelength filters and actively triggered filters are: the ability to filter a multiple number of laser lines simultaneously without hindering operation of the optical systems; the freedom from spectral and chromatic distortions that might be associated with fixed filters; and the fast response time required to protect against pulse lasers which is not achievable with actively triggered filters. The major obstacle to achieving useful optical power limiters is the high switching threshold and low damage threshold associated with most current nonlinear optical materials. Research on nonlinear optical organic and polymeric materials which potentially have high optical nonlinearity is underway to overcome these material limitations.

Optical power limiting behavior, or decrease in transmittance with increase in incident light intensity, can arise from a number of physical phenomena. These phenomena include nonlinear absorption[1], free-carrier absorption in semiconductors[2], second harmonic generation[3], stimulated Rayleigh[4], Brillouin[5], or Raman[6] scattering, reverse saturable absorption[7], and near-resonance laser produced plasma in an atomic vapor[8]. Optical power limiting has also been demonstrated based on nonlinear refractive mechanism by focusing a low-powered radiation onto an external aperture[9] so that the transmitted light through the aperture is dependent on the incident laser power. Pulse broadening has been produced using nonlinear absorption in semiconductors and active feedback techniques. Pulse steepening by saturable absorption is used in most mode-locking lasers. In this paper, we report observation of a remarkably sharp intensity limiting behavior in soluble organic and polymeric materials <u>without</u> using an aperture. This intensity limiting behavior is attributed to nonlinear absorption. We have also observed pulse broadening and pulse steepening, which we believe are due to self-defocusing and focusing effects arising from refractive nonlinearities.

2. NONLINEAR OPTICAL POLYMERS AND MODEL COMPOUNDS

Current interest in the nonlinear optical (NLO) properties of organic solids and polymers for potential applications in various optical devices has been simulated by the early observations of very large nonresonant nonlinear susceptibilities in some highly conjugated molecules[10-12] and recent demonstration of both second-order and third-order nonlinearities in many organic solids and in polydiacetylenes[13-15]. Since our primary interest is in intensity limiting phenomena, our focus here will be organic materials exhibiting third-order nonlinear optical effects. The conjugated polyene model compound β-carotene, whose molecular structure is shown in Figure 1, was among the first organic NLO materials investigated by Hermann and Ducuing[10-11] who showed that β-carotene glass has a third-order molecular hyperpolarizability $\gamma = 8 \times 10^{-33}$ esu; this translates to $\chi^3 \sim 10^{-11}$ esu. The origin of this large NLO effect in β-carotene has been attributed to electronic motions in the delocalized π-electron system of the eleven conjugated double bonds. In fact, the dominant role of the π-electrons of organic materials in their NLO properties is now well accepted[10-16]; thus, if the nonlinear optical response is purely electronic in origin, time scales of the order of picosecond or less are anticipated. However, depending on the molecular structure, the state of the sample, degree of molecular orientation in a macroscopic sample, and other factors, the nonlinear optical response of organic and polymeric materials may have both an electronic component and other mechanisms.

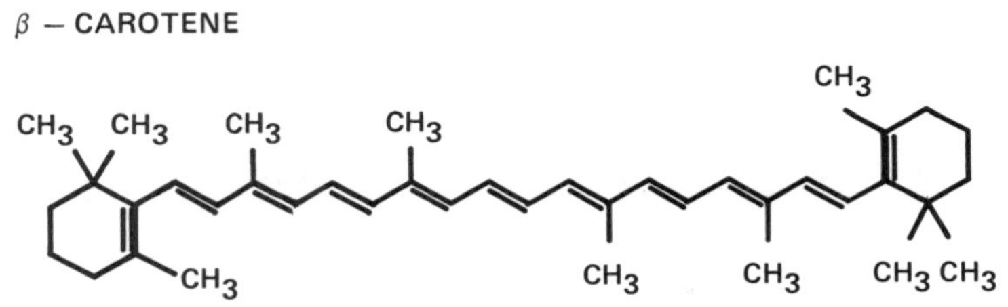

β — CAROTENE

Figure 1. Molecular Structure of β-carotene

Third-order nonlinear optical effects, including intensity limiting behavior, were studied in three different classes of organic model compounds and polymers: (1) β-carotene; (2) polyphenazine model compounds; and (3) heteroaromatic polymers. Samples of β-carotene were purchased from Eastman Kodak and used as received.

The monomer 5,10-diphenylphenazine (DPP) and dimer biphenylene-bridged phenazine (BBP) whose molecular structures are shown in Figure 2 were synthesized and characterized as model compounds of the liquid crystalline polymers of 5,10-disubstituted phenazines that are bridged by parasubstituted phenylene groups. One unique feature of these compounds is their propensity to oxidize to relatively stable radical cations. Both DPP and BBP were dissolved in dichloroacetic acid (DCA) to oxidize them prior to characterization of NLO properties. The predicted nonlinear response mechanism involved charge-transfer of the radical cation across the bridge group to produce a separate structure of localized exciton having a considerable dipole moment, as shown in Figure 3. Unfortunately, due to steric hindrance between the ortho hydrogens on the phenyl group and the 1, 9, 6, 4 phenazine hydrogens, only a small orbital mixing through the phenyl bridge group is achieved with the dimer structure. Hence, details of such charge-transfer induced nonlinear response mechanism could not be fully evaluated based on the model compounds.

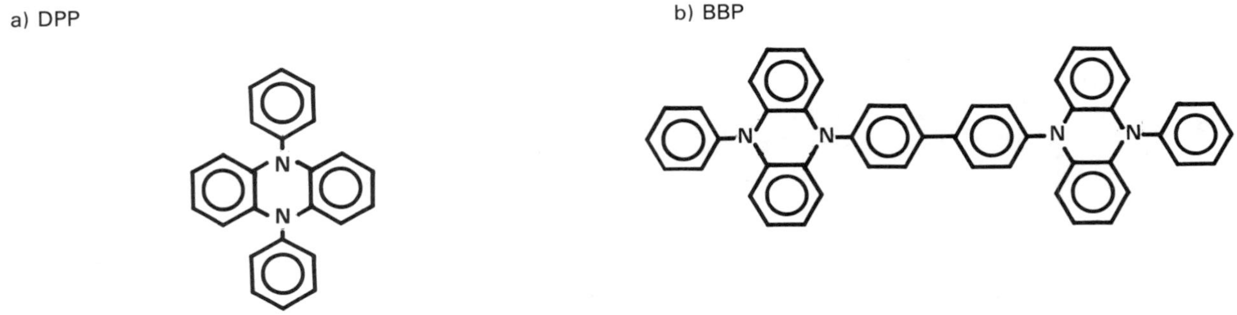

a) DPP

b) BBP

Figure 2. Molecular Structure of DPP and BBP

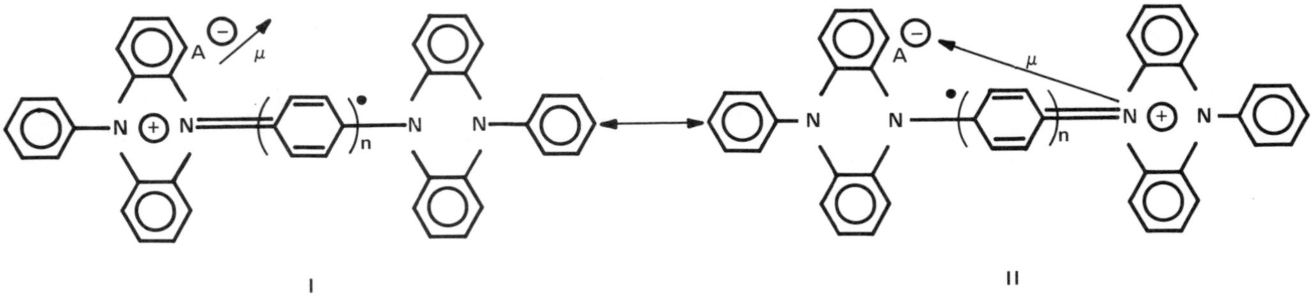

I

II

Figure 3. Nonlinear Optical Mechanism via Charge-Transfer of Radical Cation in PPZ.

The third class of organic materials investigated consists of the heteroaromatic polymers whose structures are shown in Figure 4, including the polythiophenes (X = S) and polypyrroles (X = N-H). A detailed description of the synthesis and structural characterization of these polymers will appear elsewhere[17]. The names of those members of this class of polymers described here are summarized in Table 1. According to Agrawal, Cojan, and Flytzanis (ACF) band model theory of third-order NLO effects in organic and polymeric materials, χ^3 is predicted to scale as the 6th power of the degree of π-electron conjugation[16]. One unique feature of these polymers is the flexibility of molecular engineering, due to the various X and R groups, possible substitutions at 3,4 positions on the ring, and the degree of π-electron conjugation along the polymer backbone, which should allow eventual structure-property correlation in nonlinear optical polymers.

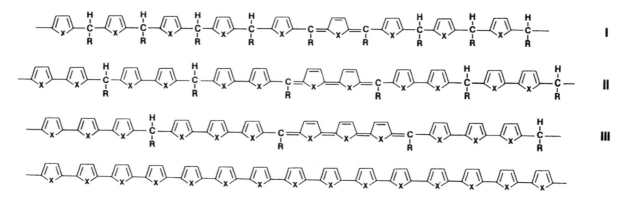

Figure 4. Chain Structures of the Heteroaromatic Polymers with Increasing Degree of Conjugation.

Table 1. Structure, Abbreviation, and Name of Heteroaromatic Polymers.

Stereoregular Heteroaromatic Polymer	Abbreviation	Structure
Poly (2,5-thiophenediyl benzylidene)	PTB	Figure 4 (I):X = S, R = Ph.
Poly (2,5-thiophenediyl p-nitrobenzylidene)	PTNB	Figure 4 (I):X = S, R = Ph-NO$_2$.
Poly (2,5-pyrrolediyl benzylidene)	PPB	Figure 4 (I):X = N − H, R = Ph.
Poly (α-[5,5'-bithiophenediyl] benzylidene)	PBTB	Figure 4 (II):X = S, R = Ph.
Poly (α-[5,5'-bithiophenediyl] p-acetoxybenzylidene)	PBTAB	Figure 4 (II):X = S, R = p-Acetoxy-Ph.
Poly (α-[5,5″-terthiophenediyl] benzylidene)	PTTB	Figure 4 (III):X = S, R = Ph.

DMF = dimethylformamide
DCM = dichloromethane
THF = tetrahydrofuran
Ph = phenyl

3. NONLINEAR OPTICAL CHARACTERIZATION TECHNIQUES

3.1 Experimental Technique

A number of third-order nonlinear optical phenomena have been used by different research groups to characterize the nonlinear optical properties in materials, including four-wave mixing[18], third-harmonic generation[19], electric field induced second-harmonic generation[20], and nonlinear transmission measurement[21] techniques. We have chosen nonlinear transmission measurement technique because it is possible to measure the nonlinear absorption and refractive indices directly, based on a variety of nonlinear optical effects of wave propagation through a nonlinear medium. In particular, we have used intensity limiting transmission[21] and temporal pulse steepening[22] effects in nonlinear wave propagation in deducing the nonlinear optical indices. For nonlinear absorption measurement, we have used two different nanosecond and a picosecond Nd:YAG laser at 1064 nm for characterization. The lasers used include a Quanta Ray (DCR model) electro-optical Q-switched laser, a Quantronix (114 model) acousto-optical Q-switched laser, and a Quantel 10-20 Hz, 35-picosecond, mode-lock laser. For nonlinear refraction measurement via the temporal pulse steepening phenomenon, we have used only a nanosecond laser, due to the difficulties in resolving the temporal pulse width of a picosecond laser.

The basic experimental set-up for nonlinear absorption and refraction measurement is shown in Figure 5. In this set-up, the intensity of the Nd:YAG laser incident on the sample (1 cm thick liquid cell or ~10 μm thin solid films) was controlled by an attenuator arrangement using a half-wave plate followed by a linear polarizer. The incident and the transmitted energy monitors were calibrated and properly matched pyroelectric detectors, such that their output, when fed into a ratiometer, would directly measure the transmittance through the sample. Typically, an average of 10 to 100 ratios were recorded for each intensity setting to minimize pulse-to-pulse variations. The pulse width of the incident and transmitted beams were monitored by two fast Si PIN diodes whose outputs were recorded on a fast storage scope with resolution better than 1 ns. The irradiation on the sample could be coarsely adjusted by using lenses with different focal lengths, which gave different beam spot diameters at the sample.

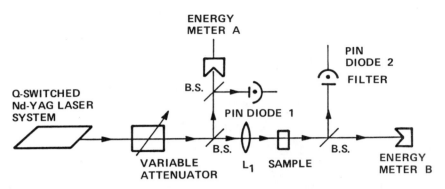

Figure 5. Experimental Set Up for Nonlinear Optical Measurements.

In the nonlinear absorption experiment, detectors with large area (5 cm^2) were used to collect all the energy transmitted such that the measured transmittance would be insensitive to any diffraction effects which could arise from refractive nonlinearity of the sample. Also, all data obtained in solution samples were normalized to that of the pure solvent so that all reported nonlinearities are intrinsic to the organic compounds or polymers.

3.2 Analytical Techniques

The theoretical model used in our analysis of experimental data is based on nonlinear absorption and self-focusing/defocusing phenomena. Using the thin medium approximation, and to the lowest order terms of the nonlinear process, it has been shown[21] that the transmittance due to nonlinear absorption is given by

$$T = \frac{2\alpha_1}{I_o \pi \alpha_2 (\exp[\alpha_1 L] - 1)} \int_0^\infty dx \ln[1 + q(O,L) \exp(-x^2)] \tag{1}$$

where

$$q(r,L) = \alpha_2 I(r,O) [1 - \exp(-\alpha_1 L)]/\alpha_1 \tag{2}$$

$$I(r,O) = I_o \exp(-r^2) \tag{3}$$

where r is the radial coordinate of the beam inside the nonlinear medium, $n_{o,2}$ are the linear and nonlinear refractive indices, $\alpha_{1,2}$ are the linear and nonlinear absorption coefficients, respectively, L is the thickness of the nonlinear medium, k is the radiation wave number, and I_o is the peak power density of the incident beam.

The first-order approximation in α_2 of equation 1 can be further simplified by expanding $\ln[1 + q \exp(-x^2)]$. This gives a simple relationship between the transmittance and the nonlinear absorption coefficient α_2:

$$T = \exp(-\alpha_1 L) \times [1 - \alpha_2 I_o (1 - \exp(-\alpha_1 L))/2\sqrt{2}\alpha_1] \tag{4a}$$

or

$$\ln T = \alpha_2 I_o [1 - \exp(-\alpha_1 L)]/2\sqrt{2}\alpha_1 \tag{4b}$$

Thus, to a first-order approximation, the nonlinear absorption coefficient is given by the slope of the natural logarithm of the transmittance vs peak energy density. In our data analysis, we used equation 4b as the first estimate for α_2, and then used equation 1 to obtain a more accurate value for α_2.

For the nonlinear refractive index derivation, we use a theoretical approach based on the time-dependent analysis of the on-axis pulse width. It has been shown that [22,23]

$$I(z,t) = I_o (t - z/v) [1 - (z/z_f)^2]^{-1} \tag{5}$$

where

$$z_f = ka_o^2/(I_o' - 1)^{1/2} \text{ is the transient optical-induced focal length in the medium and}$$

$$I_o' = (n_2 k^2 a_o^2/n_o) I_o \tag{6}$$

where a_o is the incident beam radius.

Assuming the pulse shape of the incident beam is Lorentzian with a Full Width Half Maximum (FWHM) = t_o such that

$$I_o(t) = I_o[1 + (t/t_o)^2]^{-1} \tag{7}$$

then the half-width τ of the on-axis pulse at L is given by

$$\tau^2/t_o^2 = [1 - (L/z_f)^2]/(1 + L')$$ (8)

where L' is the thickness of the sample divided by the characteristic length $R_d = \frac{1}{2} ka_o^2$. Thus, by measuring the FWHM of the incident and transmitted beam, one would be able to obtain z_f and hence, n_2. A negative n_2 would produce an imaginary z_f, so that one would observe a pulse-broadening effect, whereas a positive n_2 such that $I_o' > 1$ (when self-focusing occurs) would produce a positive z_f, in which case one would observe a pulse-steepening effect.

4. RESULTS AND DISCUSSION

4.1 Nonlinear Absorption and Optical Limiting

The nonlinear absorption coefficient (α_2) data obtained with the three different lasers are collected in Table 2 for the five heteroaromatic polymers and three model compounds investigated. The largest absorptive nonlinearity was observed in PBTB, which has α_2 value of about 10^{-7} cm/W with nanosecond pulses. This value of α_2 is about one order of magnitude larger than that observed in β-carotene. It should be noted that the nonlinear absorption coefficient reported in Table 2 is an implicit function of the solution concentration; thus, the α_2 values are best compared when normalized to molar concentrations.

Table 2. Summary of Nonlinear Material Properties Measured

Polymer Or Compound	Polymer Type	Nominal Conc. (molar)	Nanosecond Results				Picosecond Result $\alpha_2(3)^*$ (cm/W)
			N.L. Threshold (W/cm²)	$\alpha_2(1)^*$ (cm/W)	$\alpha_2(2)^*$ (cm/W)	n_2 (cm²/W)	
PTB	HP (I)	4.4×10^{-3}	$>6\times10^8$	1.5×10^{-9}	1.2×10^{-8}	-7.4×10^{-13}	1.7×10^{-11}
PPB	HP (I)	3.2×10^{-3}	10^8	2.6×10^{-8}	2.6×10^{-8}	+	3.1×10^{-11}
PBTB	HP (II)	2.0×10^{-3}	10^7	3.2×10^{-7}	5.8×10^{-7}	1.5×10^{-11}	9.4×10^{-11}
PTTB-2	HP (III)	1.5×10^{-3}	$>10^8$	1.2×10^{-7}	+	+	+
PBTAB	HP (II)	1.5×10^{-3}	2×10^8	1.0×10^{-7}	6.9×10^{-9}	-6.9×10^{-14}	1.9×10^{-11}
βC	PA	9.4×10^{-3}	3×10^8	4.0×10^{-9}	2.6×10^{-8}	1.4×10^{-12}	2.2×10^{-10}
DPP	PPZ	10^{-2}	10^8	1.2×10^{-7}	7.4×10^{-8}	3.0×10^{-13}	Bleaching
BBP	PPZ	10^{-2}	2×10^8	1.6×10^{-8}	Bleaching	4.0×10^{-13}	Bleaching

HP = Heteroaromatic Polymers
PA = Polyacetylene
PPZ = Polyphenazines

* $\alpha_2(1)$ is the Quanta Ray Result
$\alpha_2(2)$ is the Quantronix Result
$\alpha_2(3)$ is the Quantal Result

+Not measured

The bleaching noted in Table 2 for DPP and BBP may be due to chemical degradation of the samples prior to NLO measurements rather than a photochemical mechanism since some discoloration of the originally green oxidized solution samples of DPP and BBP was observed after many months of their preparation.

The nonlinear absorption coefficient measured with 35 picosecond laser pulses is also shown in Table 2. At this switching speed it is interesting that both PBTB and β-carotene have the same value of α_2, about 10^{-10} cm/W, whereas PBTB exhibited an order of magnitude higher α_2 with nanosecond pulses. Also, we note that α_2 values in the picosecond range are one to three orders of magnitude smaller than values in the nanosecond range, the smallest difference being in PBTB and β-carotene. If the observed nonlinearities in the picosecond range are due to a pure electronic mechanism, the present results suggest that alternate mechanisms, such as thermal or resonance effects, may have enhanced the nonlinear absorption with nanosecond pulses. To further understand the origin of the difference between nanosecond and picosecond nonlinear absorption coefficients, we examined α_2 data for the possible dependence on energy density and pulse width. Comparison of α_2 as a function of energy density at fixed pulse width showed a constant value. However, comparison of α_2 at different pulse widths with the same range of energy density showed identical orders of magnitude difference (Table 2). Hence, since the magnitude of the absorptive nonlinearity does not depend on energy density, the nonlinearity is probably not thermal in origin. The response time of the nonlinear mechanism is clearly longer than the 35 picosecond pulse width of the laser.

The absorptive nonlinearities in thin solid films (< 10 μm) of the samples in Table 2 were much harder to detect because the damage threshold was very close to the nonlinear absorption threshold. However, we detected absorption increase for succeeding pulses before the films were damaged.

Figures 6 to 8 show the optical power limiting behavior in β-carotene, BBP model compound, and PBTB respectively. Similar results of optical power limiting were obtained in all the model compounds and polymers of Tables 1 and 2, except PTNB which did not show any measureable NLO effects. The nonlinear threshold for intensity limiting, determined from Figures 6 to 8 and similar results as the intercept between nonlinear and linear transmission, is shown in Table 2. The smallest nonlinear threshold (10 MW/cm²) was observed in PBTB which also has the highest absorptive nonlinearity among the materials reported here. We attribute the observed optical power limiting behavior to nonlinear absorption in view of our experimental configuration which was designed to eliminate contributions from diffraction effects. This is also suggested from a comparison of the measured values of both α_2 and n_2 in Table 2. Optical power limiting of the same or better than β-carotene was observed in samples (e.g. BBP) with one order of magnitude smaller n_2 values as β-carotene. However, some contribution of nonlinear refraction to the observed optical power limiting cannot be totally ruled out since the materials also have large values of n_2 as described subsequently in section 4.2. If the power limiting behavior somehow depends on n_2, it will probably be indirectly through the ratio α_2/n_2 as the theoretical results of Hermann[24] suggest for the case of an optical power limiter model with an aperture.

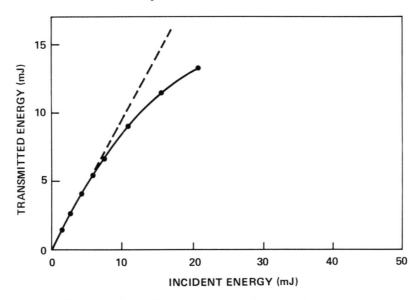

Figure 6. Optical Power Limiting Behavior in β-carotene.

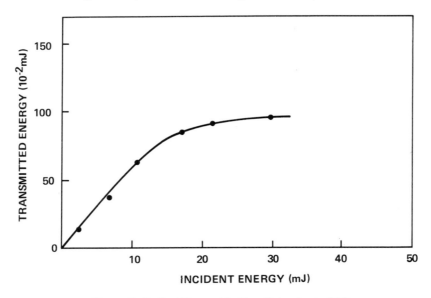

Figure 7. Optical Power Limiting Behavior in BBP.

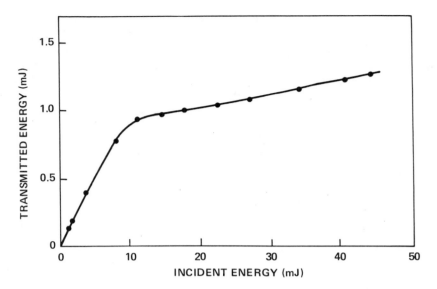

Figure 8. Optical Power Limiting Behavior in PBTB.

4.2 Nonlinear Refraction and Pulse Steepening/Broadening Effects

As pointed out in Section 3.2, the temporal pulse shape of the transmitted pulse through nonlinear optical media is predicted from self-focusing/defocusing theory to be steepened or broadened, depending on the sign of the nonlinear refractive index n_2. Indeed, by recording the transmitted pulse shape and the incident pulse shape for different incident intensity, we observed a definite trend in pulse steepening for solution samples of PBTB, β-carotene, DPP, and BBP, and pulse broadening for solution samples of PTB and PBTAB. A comparison of the incident pulse and transmitted pulse shapes showing pulse steepening and broadening effects in PBTB and PBTAB solutions, respectively, is shown in Figure 9. Figure 10 (a) shows the increase in pulse steepening in PBTB, from 65 ns to 52 ns pulse width, with increase in incident intensity from 5×10^6 W/cm^2 to 5×10^7 W/cm^2. Similar pulse steepening phenomena were observed in β-carotene, DPP, and BBP. In PBTAB, the increase in pulse broadening (from 88 ns to 99 ns pulse width) as a function of incident intensity (from 3×10^7 W/cm^2 to 7.4×10^7 W/cm^2) is shown in Figure 10 (b). These results are in excellent qualitative agreement with predictions of self-focusing/defocusing theory.

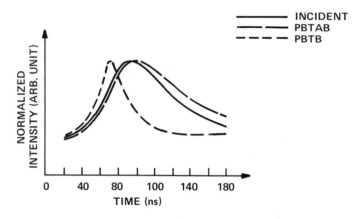

Figure 9. A Comparison of the Incident and Transmitted Pulse Shapes through PBTB and PBTAB Solutions.

a)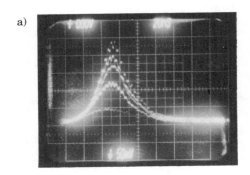

b)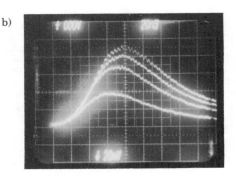

Figure 10. Transmitted Pulse Shapes as a Function of Incident Intensity.

a) Transmitted pulse shapes through PBTB (upper to lower traces correspond to incident peak power densities: 5.8×10^7 W/cm^2; 4.3×10^7 W/cm^2; 2.98×10^7 W/cm^2; 1.57×10^7 W/cm^2; 5.41×10^6 W/cm^2)

b) Transmitted pulse shapes through PBTAB (upper to lower traces correspond to incident peak power densities: 7.4×10^7 W/cm^2; 6.56×10^7 W/cm^2; 5.62×10^7 W/cm^2; 3×10^7 W/cm^2)

Using the analytical technique outlined in Section 3.2, n_2 for the polymers and model compounds was extracted and is displayed in Table 2. Note that the sign of n_2 for PBTB, β-carotene, DPP, and BBP is positive and therefore, should lead to self-focusing, whereas the sign of n_2 for PTB and PBTAB is negative, which should lead to self-defocusing of the laser pulse as observed. The values of n_2 in Table 2 range from 7×10^{-14} cm^2/W to 1.5×10^{-11} cm^2/W, with the highest value obtained in PBTB. Comparing the n_2 values of PBTB and β-carotene, we find that the nonlinearity in PBTB is larger by one order of magnitude. In fact, the refractive nonlinearity in PBTB is of the same order of magnitude as the best organic polymer (polydiacetylene) reported to date.[13-15] By normalizing the n_2 value measured in our β-carotene solution to 1 molar concentration, we obtain an n_2 value for β-carotene in good agreement with that reported by Hermann, et. al for β-carotene glass.[10]

5. CONCLUSIONS

The third-order nonlinear optical properties of several organic model compounds and polymers have been investigated, including measurements of the nonlinear absorption coefficient (α_2), nonlinear index of refraction (n_2), and demonstration of optical power limiting behavior, pulse steepening, and pulse broadening effects. The largest third-order nonlinear optical coefficients were observed in a solution ($\sim 10^{-3}$ molar) of the heteroaromatic polymer PBTB, with α_2 of about 10^{-7} cm/W and n_2 of about 10^{-11} cm^2/W. To our knowledge, our nonlinear absorption coefficients are the first reported for organic polymers. The observed value of n_2 in PBTB is one order of magnitude higher than in β-carotene and is of the same order of magnitude as the best organic NLO polymer (polydiacetylene-PTS) reported to date.

A novel method for directly measuring the nonlinear index of refraction (n_2) in the nanosecond regime from pulse steepening and pulse broadening effects is shown to be a sensitive and simple way of evaluating both the sign and magnitude of n_2.

Overall, observation of very sharp intensity limiting behavior in several new nonlinear optical polymers demonstrates the basic feasibility of an optical power limiter using these organic materials in the nanosecond regime. This holds promise for applications in passive optical limiting and regulating devices.

ACKNOWLEDGEMENTS

This work was supported in part by the Naval Air Development Center, Warminster, Pennsylvania under Contract N62269-85-C-0718 and Army Natick RD&E Center, Natick, Massachusetts under Contract DAAK 60-84-C-0044.

REFERENCES

1. A. Hordvik, IEEE J. Quantum Electron QE-6, 199 (1970).
2. W.W. Piper and D.T.F. Marple, J. Appl. Phys. 32, 2237 (1961).
3. I.C. Khoo and S.L. Zuang, Appl. Phy. Lett. 37, 3 (1980).
   I.C. Khoo, Phys. Rev. A, 27, 2747 (1938.)
4. R.H. Pantell and H. Warszawski, Appl. Phys. Lett. 11, 213 (1967).
5. A.J. Alcock and C. DeMichelis, Appl. Phys. Lett. 11, 42 (1967).
6. J.R. Murray, J. Goldhar, D. Eimerl, and A. Szoke, IEEE J. Quantum Electron QE-15, 342 (1979).
7. D.J. Harter, M.L. Shand, and T.B. Band, J. Appl. Phys. 56, 865 (1984).
8. A.C. Tam, Appl. Phys. Lett. 35, 683 (1979).
9. M.J. Soileau, W.E. Williams, and E.W. Van Stryland, IEEE J. Quantum Electron QE-19, 731 (1983).
10. J.P. Hermann, D. Richard, and J. Ducuing, Appl. Phys. Lett. 23, 178 (1973).
11. J.P. Hermann and J. Ducuing, J. Appl. Phys. 45, 5100 (1974).
12. K.C. Rustangi and J. Ducuing, Opt. Commun. 10, 258 (1974).
13. C. Sauteret, J.P. Hermann, R. Frey, F. Pradere, J. Ducuing, R.H. Baughman, and R.R. Chance, Phys. Rev. Lett. 38, 956 (1976).
14. D.J. Williams, editor, Nonlinear Optical Properties of Organic and Polymeric Materials, ACS Symposium Series No. 233, Am. Chem. Soc., Washington, D.C., 1983.
15. G.J. Bjorklund, et al., Appl. Optics 26, 227 (1987).

16. (a) G.P. Argrawal, C. Cojan, and C. Flytzanis, <u>Phys. Rev. B 17</u>, 776 (1978).
    (b) C. Flytzanis, C. Cojan, and G.P. Agrawal, <u>Nuovo Cimento</u> (Italy) <u>39B</u>, 488 (1977).
17. S.A. Jenekhe, <u>Macromolecules</u>, submitted.
18. A.M. Dennis, W. Blau, and D.J. Bradley, <u>Appl. Phys. Lett. 47</u>, 200 (1985)
19. G.R. Meridith, B. Buchalter, and C. Hanzlik, <u>J. Chem. Phys. 78</u>, 1543 (1983).
20. A.F. Garito, K.D. Singer, and C.C. Teng, in: ref. 14, pp. 1 ff (1983); G.R. Meredith, Ibid, pp. 27 ff (1983).
21. J.A. Hermann, <u>J. Opt. Soc. Am B. 1</u>, 729 (1980); E.W. Van Stryland, H. Vanharxeale, M.A. Woodall, M.J. Soileau, A.L. Smirl, S. Guha, and T.F. Boggess, <u>Opt. Eng. 24</u>, 613 (1985).
22. J.H. Marburger and W.G. Wagner, <u>IEEE J. Quantum Electron QE-3</u>, 415 (1967).
23. T.R. Shen, <u>Principles of Nonlinear Optics</u>, Wiley: New York, 1984, Chapt. 17.
24. J.A. Hermann, <u>Optical Acta 32</u>, 541 (1985).

Polymeric non-linear optical waveguides.

J. Brettle, N. Carr, R. Glenn, M. Goodwin and C. Trundle.

Plessey Research Ltd., Allen Clark Research Centre,
Caswell, Towcester, Northants., England. NN12 8EQ.

ABSTRACT

A particularly useful format for organic non-linear optical materials is that of a planar or stripe waveguide so that the materials can be utilised in integrated optical devices. The materials then have requirements placed upon them in addition to the need for high χ^2 or χ^3 coefficients: they must also be depositable in films of well controlled thickness and low scatter to enable low loss waveguiding to be performed over a sufficient distance to achieve non-linear interactions.

Three approaches to achieving this in polymeric systems are described:

1. Non-linear dopant/film forming polymer/solvent systems where high levels of dopant can be achieved without phase separation in solvent cast, spun polymer films.
2. Solvent assisted indiffusion of non-linear dopant molecules into a polymer matrix material to produce slab or strip waveguides.
3. Deposition of novel non-linear optical polymers using the Langmuir Blodgett technique. Multilayer films have been built up to a sufficient thickness to achieve low loss waveguiding.

For each of these approaches the choice of non-linear and/or polymer material will be discussed together with a description of the fabrication technique and optical properties of the finished structure. The applicability and suitability of these approaches to the fabrication of non-linear optical devices will be compared.

1. INTRODUCTION

Organic materials have been recognised as being attractive for non-linear optical applications for some time because of their ability, potentially at least, to combine high non-linear coefficients, performance over a wide wavelength range, resistance to laser damage and high operating speed. These potential advantages must be weighed against their disadvantages such as poorer environmental stability compared to inorganic materials and the problems of integrating organic materials with their limited temperature stability into well established process technologies which were originally developed for inorganic materials.

Perhaps the most attractive format for an organic non-linear optical material is that of a thin polymer film. This is because:

(i) such films are potentially easy to deposit e.g. by spinning, dipping or other coating technologies,

(ii) the films can be used as waveguides in integrated optical devices. The use of a waveguide format is attractive in that it relaxes some of the phase matching restraints relating to bulk devices and makes interconnection and integration of devices much easier.

It seems likely that bulk organic single crystal devices will find only limited use in the major application fields of telecommunications and optical data processing. In addition, because of the much more difficult technology involved in crystalline film growth compared to polymer film deposition the use of thin crystalline films will only develop if sufficiently high coefficients cannot be achieved in polymer films.

For these reasons we have concentrated our efforts on polymer films. Incorporation of high concentrations of non-linear organic moieties in a polymer matrix material may be achieved by a variety of techniques such as:

(a) dissolution of non-linear dopant in molten thermoplastic polymer and subsequent solidification

(b) addition of dopant to a reactively curing thermosetting polymer

(c) achieving an intimate mixture of dopant and matrix molecules by casting them together from a solution in a common solvent

(d)   diffusing dopant molecules into a solid matrix polymer

(e)   making the dopant an intrinsic part of the matrix polymer by grafting it on as a side chain or part of the main chain.

Each of these approaches has its advantages and drawbacks, in particular incorporation in molten thermoplastic implies using a low melting point (and hence a low Tg) polymer if one is not to be limited to dopant molecules stable at high temperatures; incorporation in a reactively cured system requires a chemically resistant dopant if it is not to be degraded by the free radicals present in the curing reaction. For these reasons approaches c,d and e were investigated.

2. SOLVENT CASTING OF DOPED FILMS

This approach has been used by other workers e.g. with PMMA (polymethyl methacrylate)[1] or PVDF (polyvinylidene difluoride)[2] as a matrix material. We have worked with a variety of polymers, results on PMMA and a novolak (cresol-formaldehyde) resin are discussed below.

The solvent casting was performed using a conventional microlithographic spinning technique. Solvents need to be choosen with care to achieve good spinning properties; it should be appreciated that the solubility of the dopant in the polymer solution may be different from that in the polymer alone as a solid solution, this can lead to unwanted crystallisation of the dopant with optical scattering from the resultant two phase film. It is also possible that solubility of the dopant is greater in the dry film such that spinning from a warm solution (to increase saturated dopant concentration) may result in a single phase film at ambient temperatures.

As an example of this approach 35g of low molecular weight PMMA were dissolved in 100ml of 2-ethoxyethyl acetate and 7g of MNA (2-methyl-4-nitroaniline) were added. The resulting solution was used to spin-coat a BK7 glass substrate at 5000 rpm for 45 sec. The sample was then baked at 50^0C overnight to remove solvent.

The linear optical waveguide properties of these films were assessed by evaluation of propagation modes measured by prism coupling at 633nm, this yielded film thickness and refractive index. Monomode guides, typically 2-3μm thick could be produced with refractive index increases over the substrate (n=1.515 at 633nm) of 0.02 to 0.03. Waveguide propagation losses were determined by scanning a fibre bundle along a propagating waveguide mode and monitoring the scattered light as a function of distance. Losses of 0.75 ± 0.25 dB cm could be achieved.

Prism/waveguide[3] coupling was used to investigate the non-linear properties of the guide. This technique uses the change in refractive index caused by a high intensity optical imput to change the synchronous coupling condition and thereby bring about a reduction in prism/coupler efficiency. A characteristic of this effect is that full coupler efficiency can be recovered at high power by altering the incident beam angle to restore synchronicity. Fig 1 shows this effect for a waveguide excitation in the TE0 mode at 1.06μm, high power pulses of 25ns duration were provided by a Q-switched Nd-YAG laser. The reduction in coupler efficiency and its restoration by imput beam realignment can be observed indicating a non-linear refractive index capable of responding to the laser fast pulse rise time of < 5 ns.

Quadratic electro-optic modulation has also been demonstrated in these films using a slab waveguide of 17% MNA in PMMA in one arm of a Mach-Zender interferometer, a χ^3 value of 6 x 10^{-20} CmV^{-3} was measured. The fact that this is somewhat lower that calculable from values for bulk MNA is due to the less than perfect electrical and optical field overlap in the slab guide, stripe guides produced using the in-diffusion technique described below would be useful here.

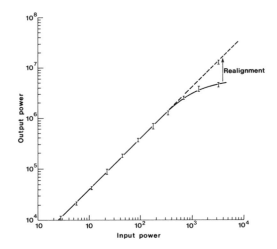

Fig. 1 NON-LINEAR PRISM COUPLING OF A SOLVENT CAST MNA/PMMA WAVEGUIDE AT 1.06µm

More recently we have investigated the use of a novolak resin as host polymer. Novolaks are a family of cresol-formaldehyde condensation polymers, they were choosen for the following reasons:

(i)   these resins are known to be good "film formers" from their use in lithography and the coatings industry

(ii)   their polar nature suggests that they should be compatible with, and therefore a good solvent for, polar chromophoric dopants

(iii) the polymers are amorphous, the absence of a two-phase structure is advantageous in minimising intrinsic scattering effects in the films.

(iv)   the use of novolaks as the base resin in positive resists suggests that they should be processible using microcircuit fabrication techniques and that by adding a second photoactive material to the novolak a photo-patternable film could be produced which could be a useful waveguide fabrication technique.

(v)   they are capable of patterning to produce stripe waveguides using the solvent in-diffusion technique discussed below.

Higher concentrations of dopant have been achieved in novolak than PMMA e.g. up to 40 w/o polymer in the case of MNA, also the novolak is much easier to in-diffuse than PMMA. Unfortunately the novolak waveguides produced so far are of poorer quality than the PMMA guides, with losses of several dB/cm, we believe this to be due to impurities in the novolak; work is continuing to improve these.

3. DOPING OF FILMS BY THE SAID (SOLVENT ASSISTED IN-DIFFUSION) TECHNIQUE[4]

In this method a polymeric surface is in-diffused with an organic dopant molecule by immersing the polymer in a hot solution of the dopant. This method is commonly used in the textile industry to dye artificial fibres and in opthalmics to tint plastic lenses. A surprisingly large concentration of solute can be in-diffused by this method if the conditions and the solvent are choosen correctly; it is necessary that the solvent is stable at the high temperatures used and that the solubility of the dopant in it is not too great. It is necessary that the dopant molecule is more soluble in the polymer matrix than in the solvent to achieve a thermodynamic driving force for the diffusion. For these reasons a solvent in which the dopant is sparingly soluble is choosen and an excess of solid dopant is retained as a dispersion in the diffusion vessel to maintain saturation of the solution as the dopant is transferred to the polymer, this procedure is known in the textile industry as disperse dyeing. We have found fluorocarbon solvents particularly useful for the dopants which we have wanted to diffuse.

Typically the solvent is heated to a temperature which is determined by:

(i)   solubility of the dopant in both solvent and polymer

(ii)   melting point of the dopant

(iii)   sublimation temperature of the dopant

A temperature lower than the melting point must be used or the molten dopant will adhere to the polymer causing uneven doping, the temperature must also be lower than the sublimation temperature to prevent loss of the dopant from solution which would reduce its concentration (this is a particular problem with nitroanilines).

Table 1 gives details of dopant materials, solvents, experimental conditions and optical results. A variety of polymers have been investigated but CR39 (poly allyldiglycol carbonate from PPG), a highly cross-linked polymer used for opthalmic lenses, has proved particularly successful.

TABLE 1

Fabrication parameters for waveguides prepared by solvent-assisted indiffusion into CR39 polymer substrates

Material (see text)	*Solvent	Temperature (^{0}C)	Immersion time(s)	Surface Δn	**Number of Modes
DAN	FC40	155	15	0.001	2
o-NA	FC40	155	15	0.01	5
Cl-NA	FC40	155	30	0.006	8
m-NA	FC72	56	600	0.005	2

*   Perfluorinated hydrocarbon solvents, FC series, 3M Corp.
**   Measured at 633nm wavelength

Fig 2 shows the approximately gaussian refractive index versus depth profile for a particular Cl-NA (2-chloro-4-nitroaniline) / CR39 multimode guide. Monomode guides were also produced with losses of a few dB/cm. The non-linearity of these guides has been investigated in a non-linear prism coupling experiment[3]; using an Ar ion laser at 514 nm, the response of a DAN (1-dimethyl-2-acetamido-4-nitrobenzene) / CR39 guide was similar in form to that shown in fig. 1. A large intensity dependent index of 0.5 (MWcm^{-2})$^{-1}$ could be calculated from the measurements but with an associated time constant of approximately 2ms suggesting a thermal rather than an electronic origin to the effect. Experiments using the pulsed output from a passively Q switched Nd YAG laser at 1.06 μm with an o-NA (ortho-nitroaniline) / CR39 guide gave a much lower coefficient of 10^{-7} (MWcm^{-2})$^{-1}$ but with a response time at least as fast as the laser pulse (5ns). Cl-NA gave a similar result but some optical damage was observed.

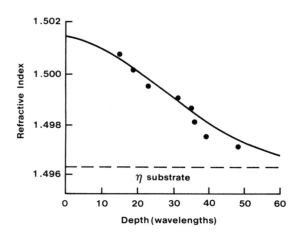

FIG. 2   DEPTH PROFILE FOR A SOLVENT ASSISTED IN-DIFFUSED GUIDE OF Cl - NA IN CR39

A major advantage or this technique is that using simple lithographic techniques channel waveguides can be defined. An aluminium layer was deposited on the polymer surface and patterned in the normal lithographic manner, the aluminium pattern then acted as as an in-situ mask allowing diffusion of dopant only through clear areas. Using this technique 3, 5 and 7μm channel waveguides and directional coupler structures have been formed in CR39. We anticipate that this method may be particularly useful in defining guides in spun films uniformly doped with a non-linear compound. In this case solvent assisted diffusion of a high index linear compound would be used to define the channel guide.

4. FILM DEPOSITION BY THE LB TECHNIQUE

This technique is capable of producing high quality, orientated films of organic materials by depositing the films one molecular layer at a time. The close control of layer thickness and molecular orientation achievable makes the technique particularly attractive for this application, the alignment of the chromophores can be achieved without the poling procedure which is often used in polymer systems for this purpose.

The technique involves formation of a compressed monolayer of material on a water subphase and the removal of this monolayer by drawing a substrate through the surface. By repeated dipping and withdrawing, multilayers can be built up on the substrate. The molecules must contain a hydrophilic "head group" and a hydrophobic "tail" to properly orientate themselves at the water surface, in "classical" LB materials this would be a carboxylic head and an alkane tail. Such fatty acid films are of poor mechanical stability however and are inactive from the non-linear standpoint.

The general problem of mechanical stability in LB films has been tackled by depositing either pre- or post- polymerised materials[5], [6]. Polydiacetylenes[6] are an example of this approach with postpolymerised materials to produce non-linear optical films. We have taken the prepolymerised approach however and have designed in collaboration with colleagues at Hull University a range of non-linear optical materials based on polysiloxane polymers. These have been sythesised at Hull University and have been shown to have the required characteristics. We believe this to be the first reported example of the use of a preformed polymer in such an application. One of these materials is shown in fig. 3. The important characteristics of this molecule are:

(i) A siloxane backbone, this overcomes the need for a long alkane tail (which is no more than a redundant diluent of the non-linear active part of the molecule once deposition has been achieved); in addition the polymer backbone results in a much more robust film structure than unlinked chains.

(ii) An alcohol hydrophilic head group, this group has been found to be particularly successful with respect to monolayer stability.

(iii) A suitable chromophore to give a large molecular hyperpolarisibility coefficient, β. This comprises an electron donor (the ether oxygen) coupled through the conjugated aromatic ring to an electron acceptor (the azo-phenyl group).

(iv) Hexamethylene spacer units between the chromophores and the siloxane backbone. The purpose of this was to prevent too great a rigidity in the polymer system which would have made deposition more difficult.

(v) Limitation of chromophoric side group substitution to approximately 50% to prevent steric crowding. Too close packing of the chromophores can again result in a rigid film, this is undesirable as a degree of flexibility is required to enable the film to survive the stress it suffers during deposition.

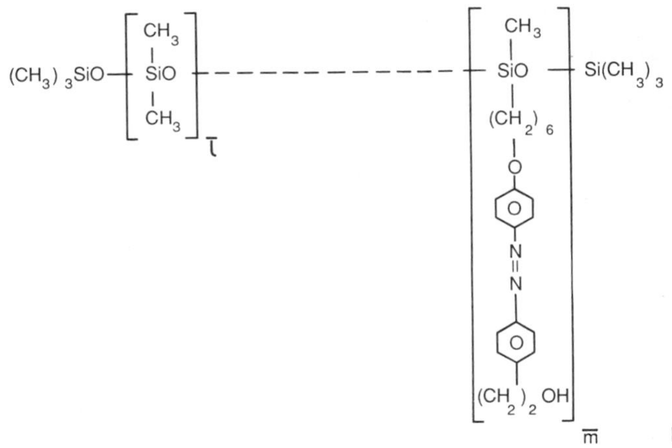

Fig. 3 SUBSTITUTED SILOXANE POLYMER USED FOR LB DEPOSITION

Experimentally, a known amount of a solution of the siloxane polymer in dichloromethane (concentration 10^{-3} M chromophores / litre was spread onto the LB subphase and compressed to form a close packed monolayer. A condensed phase is formed at a surface pressure of 20mN/m and the average area occupied per chromophoric side group is approximately 38 Å^2 at a surface pressure of 30mN/m; the observation agree well with theoretical calculations based on space filling molecular models. When the monolayer was maintained at 30 mN/m no significant change in surface area could be detected over a period of 64 hr indicating no tendency for film collapse, rearrangement, dissolution etc. A monolayer of the polymer was deposited at 50µm/sec onto a hydrophilic glass substrate and assessed for second harmonic generation. This was done by focussing linearly polarised light of 1.06µm wavelength onto the monolayer at an angle of 45^0 to the substrate and monitoring the reflected radiation for the second harmonic signal at 0.53µm. A value for the second order hyperpolarisibility of the polymer was obtained as 3.5 x 10^{-49} $C^3m^3J^{-2}$ by comparison of the SHG with that of a known material, 4-carboxy-4'-decylamethylamino-2'-methylazobenzene[7]. To obtain this value the refractive indices of the two materials were assumed to be similar and packing factors were ignored, the value must therefore be considered to be only approximate.

The ability of this system to deposit layers suitable for waveguiding was demonstrated by depositing a multilayer of over 200 layers (0.48µm thick) onto a substrate with a surface relief grating having a periodicity of 0.4467µm. The grating was used to couple 0.63µm radiation from a He-Ne laser into the waveguide modes of the film. Waveguiding was confirmed by scratching the multilayer and observing the abrupt termination of the guided light. The waveguide loss was as low as any reported for a LB film at 10 dBcm^{-1}. A much thicker film (approximately 2µm) of over 900 layers was deposited without difficulty, the average layer spacing was obtained as 22.0 Å in good agreement with the value obtained from space filling molecular models at 23 Å. The refractive indices were obtained as

$n_0 = 1.5529 \pm 0.0028$, $n_e = 1.5732 \pm 0.0044$.

No non-linear effect could be detected in these thicker films, this was expected as the multilayers were Y-type i.e. the orientations of the chromophores alternated between layers and thereby cancelled out. To obtain non-linear effects in multilayers an alternating trough system such as that used by Girling et al.[8] to construct organic superlattices will need to be used.

5. CONCLUSIONS

Three methods of producing polymeric non-linear optical waveguides have been investigated.

Rules governing the choice of conditions and polymers for solvent cast films have been developed and a novolak polymers proposed as particularly suitable. A drawback of these films is that the dopant molecules are unorientated and therefore unsuitable for χ^2 applications, we are currently developing poling of these films to overcome this problem.

Solvent assisted in-diffusion has been shown to be a particularly convenient method for defining non-linear channel waveguides both in bulk and in thin film polymer substrates, the advantage of thin film substrates being that the depth of diffusion is limited allowing high concentrations of dopant to be achieved while maintaining monomode guiding. Again the dopant molecules are not intrinsically orientated but have to rely on poling or stereochemical interaction with the polymer molecules to achieve alignment.

The LB technique is much more elegant than those above and potentially better suited to achieving the required result however the technique is complicated and requires a major effort to design and synthesis suitable molecules. The results reported above however are the first demonstration that:

(i)   preformed polymers can be used to form non-linear optical LB films

(ii)   the films can be made thick enough and of sufficiently low loss to be useful.

6. REFERENCES

(1)   J.D. Swalen, R. Santo, M. Tacke, J. Fischer, IBM J. Res. Develop., 168, (1977).

(2)   P. Pantelis, Phys. Technol.,-15-, 239, (1984).

(3)   C. Liao, G.I. Stegeman, C.T. Seaton, R.L. Shoemaker, J.D. Valera, J. Opt. Soc Am. A. -2-, 590, (1985)

(4)   M.J. Goodwin, R. Glenn, I. Bennion. Electron. Lett. -22-, 78,(1986).

(5)   R.H. Tredgold, C.S. Winter, J. Phys. D., -15-, L55, (1982).

(6)   C.W. Pitt, L.M. Walpita, Thin Solid Films,-68-, 101, (1980).

(7)   I.R. Girling, N.A. Cade, P.V. Kolinsky, J.D. Earls, G.H. Cross, I.R. Peterson. Thin Solid Films -132-, 101, (1985)

(8)   D.B. Neal, M.C. Petty, G.G. Roberts, M.M. Ahmad, W.J. Feast, I.R. Girling, N.A. Cade, P.V. Kolinsky, Electron. Lett., -22-, 460, (1986).

8. ACKNOWLEDGEMENTS

This work was performed under the Joint Opto-electronics Research Scheme (JOERS) of the U.K. Department of Trade and Industry. Particular thanks are due to collaborators at Hull University Chemistry Department, Prof. G. Gray, Drs. A.M. McRoberts, R. Marsden and R.M. Scrowston, who synthesised the materials used in the LB film work.

Organic nonlinear crystals and high power frequency conversion*

Stephan P. Velsko, Laura Davis, Francis Wang, Suzanne Monaco, and David Eimerl

University of California, Lawrence Livermore National Laboratory
P. O. Box 5508, L-490, Livermore, California 94550

ABSTRACT

We are searching for new second and third harmonic generators among the salts of organic acids and bases. We discuss the relevant properties of crystals from this group of compounds, including their nonlinear and phasematching characteristics, linear absorption, damage threshold and crystal growth. In addition, we summarize what is known concerning other nonlinear optical properties of these crystals, such as two-photon absorption, nonlinear refractive index, and stimulated Raman thresholds. A preliminary assessment is made of the potential of these materials for use in future high power, large aperture lasers such as those used for inertial confinement fusion experiments.

1. INTRODUCTION

Can organic crystals be useful for frequency converting large aperture, high power lasers, such as those used in inertial confinement fusion (ICF)?[1] We are investigating several new second and third harmonic generators based on organic moieties. These compounds come from three broad, related categories: amino acid salts, salts of carboxylic acids, and zwitterionic crystals. In this paper, we will discuss the synthetic strategy for producing these compounds, their nonlinear and phasematching properties, and a number of other properties and considerations which will determine their ultimate usefulness. Many of the considerations relevant for ICF are also relevant to other kinds of applications of crystals used for frequency doubling or mixing.

The optimization of a material for a specific frequency conversion application is, in large part, determined by some simple scaling laws.[2] Each material can be characterized by a "threshold power" (P_{th}), which is roughly the smallest peak power a diffraction limited laser pulse must have to be doubled efficiently (> 50%) by that material. A diffraction limited beam with a peak power greater than or equal to P_{th} can be doubled with high efficiency regardless of how the beam aperture is changed by telescoping--as long as the crystal length is adjusted to compensate for the change in drive intensity. The threshold power is given by

$$P_{th} = \frac{(\lambda\beta)^2}{C} \tag{1}$$

where C is proportional to the nonlinear coupling d_{eff}, and is expressed in $MW^{-1/2}$ and ß is the angular sensitivity (cm^{-1}/rad). Thus, a small angular sensitivity can be as important an attribute as a large nonlinear coupling because a larger length of crystal can be used to convert a given laser beam. By this criterion (i.e. threshold power) any crystal which has a P_{th} lower than the power available for the intended application will convert efficiently.

However, all materials which satisfy a particular threshold power criterion are in no way equivalent or equally suitable in practice. This is because, for each material, one of several undesirable processes may limit the allowable drive intensity, and hence the minimum device aperture. This, in turn will limit the minimum crystal length needed for efficient conversion. Thus, aperture limiting processes give each material a characteristic minimum volume necessary to produce efficient frequency conversion without exceeding the threshold for these undesirable effects. The cost of producing the minimum volume of each material is an unambiguous figure of merit for distinguishing materials with similar threshold powers.

The major intensity (aperture) limiting processes are: optical damage, usually characterized by a damage fluence J_D; two photon absorption; stimulated Raman or Brillouin scattering, characterized by a gain coefficient g_s; and self focusing or self phase-modulation which depends on the nonlinear refractive index, n_2. The last three processes are usually a consequence of the intrinsic chemical composition of the material, but the damage threshold is more often determined by the presence of inclusions or defects incorporated during crystal growth.[3]

---
*Work performed under the auspices of the U. S. Department of Energy by the Lawrence Livermore National Laboratory under contract number W-7405-ENG-48.

One other important material parameter is the linear absorption coefficient at the fundamental or harmonic wavelengths. Small amounts of linear absorption can become problematic when high average powers are present.[4] Clearly, for two materials with comparable absorption coefficients, the shorter crystal will absorb less total energy. In many cases, a smaller aperture will favor easier heat removal. Beyond these simple considerations, the avoidance of thermal problems depends on the thermal fracture limit and the coefficient of thermal dephasing of the material.[4] These thermal properties will not be treated in this paper, but the optical absorption characteristics of the materials under consideration will be discussed.

2. IONIC ORGANIC CRYSTALS

A large number of useful second harmonic generating crystals can be produced from molecular units composed of small conjugated ionic groups attached to organic molecules. Crystals containing these molecules can be grown from aqueous solution with various counterions, or as zwitterionic crystals. A large majority of the compounds we have examined have adequate birefringence for phasematching second and third harmonic generation of 1.064 μm light. Structural differences among these compounds leads to a variety of phasematching properties, e.g. noncritical wavelengths. Some of the crystals have nonlinearities similar to or larger than that of urea. Moreover, as a group these ionic crystals are mechanically harder and more stable in air than urea.

In fact, a number of previously discovered harmonic generators belong to this class or are related to it, and a brief summary of these kinds of crystals is contained in Ref. 5. A survey of materials with threshold powers less than or equal to half the threshold power of KDP for doubling or tripling 1.064 μm showed that nearly half the materials satisfying this criterion were in this general category.[2] However, our study represents the first systematic use of the general strategy outlined above deliberately to produce new materials in this class.

Table 1 gives the phasematching parameters for several crystals we have characterized recently: L-arginine phosphate (LAP), L-arginine fluoride (LAF), and diammonium tartrate (DAT). For comparison, the values for potassium dihydrogen phosphate (KDP) are also given. Because the crystals from this class are biaxial, and usually monoclinic, traditional means of characterizing them (e.g. wedge and prism measurements[6]) would be very slow. However, we have recently developed a method of directly determining the phasematching properties of small crystals.[7] With this method we can determine the maximum second harmonic generating efficiency of nonlinear crystals by exploring the <u>entire</u> <u>locus</u> for any phasematched process of interest. Thus, even biaxial crystals can be evaluated quickly without the need for elaborate crystal growth efforts.

Table 1. Properties of Some Ionic Organic Crystals for Doubling 1.064 μm

Material	Type	d_{eff}(pm/V)[1]	ß(cm^{-1}/rad)	P_{th}(MW)
LAP	I	1.6	6900	63
	II	1.5	4100	26
LAF	I	2.0	4900	21
	II	1.5	4100	26
DAT	I	0.86	5700	154
	II	0.59	2150	47
KDP	I	0.41[2]	4900	500
	II	0.56	2500	70

[1] Maximum d_{eff} for the given type.
[2] Based on d_{36} = 0.63 pm/V.

While organic salts do not have nonlinearities as large as those found in certain high temperature oxides, such as barium metaborate,[8] it is much easier to grow them in the sizes which will be necessary for future ICF lasers. Crystals of LAP have been grown by Cleveland Crystals Inc. under contract to LLNL to dimensions as large as 3x10x10 cm^3. Crystals of DAT and LAF have been grown by simple evaporation or cooling techniques to sizes as large as 2x2x1 cm^3 and 2x1x0.5 cm^3 respectively in our own laboratory.

The birefringence and nonlinear coefficients depend strongly on the structural details of each crystal. However, because the chemical moieties in these crystals are similar, we anticipate that many other properties which rely less critically on structural details will show marked regularities. In particular, the linear and two photon absorption properties arise from the same kinds of spectroscopic transitions from crystal to crystal. Of course, it is possible that these properties will have strong polarization dependences which vary according to structure. In the remainder of this paper, we review the general information available about this class of materials.

3. ONE PHOTON ABSORPTION

The ultraviolet edge is determined by strong Π → Π* transitions associated with the small conjugated groups. This edge is generally lower than 250 nm for the kinds of crystals we are discussing. However, LAP exhibits a significant absorption feature (1-3 %/cm) in the region between 250 and 300 nm. This is probably due to n → Π* transitions. The wavelength of such transitions has been shown to depend significantly on the molecular structure in carboxylic acids,[9] but very little information is known about their behavior in other kinds of organic salts.

Absorption in the near infrared (0.9-1.5 μm) is dominated by overtones of N-H, O-H, and C-H stretching vibrations. Extensively hydrogen bonded crystals such as LAP have overtone absorptions which lead to losses of 10-20%/cm at the Nd:YAG fundamental wavelength.[10,11] Deuteration can reduce this absorption to less than 1%/cm.[10] C-H stretching overtones are significant at 0.910 and 1.2 μm. These covalently bonded protons are not as easily replaced by deuterium, however.

4. TWO PHOTON ABSORPTION

Almost no information is available concerning two photon absorption in this class of compounds. Physical intuition suggests that the same Π → Π* transitions which determine the 1 photon UV edge are also significantly two photon active. (In fact, they must be if these transitions are to figure strongly in the $\chi^{(2)}$ response.) Thus, the "2-photon edge" is expected to begin around 400 nm, making the strength of such transitions an important issue for generating harmonics in the near ultraviolet.

5. NONLINEAR REFRACTIVE INDEX

The nonlinear index determines the threshold intensities for catastrophic self focusing. It correlates strongly with the size of the linear index and its dispersion. Measurements on LAP at LLNL indicate that n_2 is not significantly different than what would be predicted by refractive index scaling.[12] Table 2 contains n_2 data for LAP, KDP and potassium titanyl phosphate (KTP) for comparison.

Table 2. Nonlinear Refractive Index (n_2) Values for Some Frequency Doubling Crystals

Crystal	n	n_2 (x10^{13}cm^3/erg)
KDP	1.49	1.0 - 3.6[1]
LAP	1.55	1.87 - 3.04[2]
KTP	1.77	2.5 - 5.7[2]

[1] W. L. Smith, "Nonlinear Refractive Index" in Handbook of Laser Science and Technology, Marvin J. Weber, ed., CRC Press, Boca Raton, Florida (1986).
[2] R. Adair and L. Chase, LLNL, private communication.

6. STIMULATED RAMAN SCATTERING (SRS)

High frequency, narrow Raman resonances can lead to low thresholds for stimulated Raman scattering. Intense transverse SRS can cause significant energy loss or even physical damage to the nonlinear crystals.[13] Of primary concern in organic crystals are aliphatic C-H vibrations in the 2900 cm^{-1} region. These appear to have spontaneous Raman scattering intensities 2-3 times larger than other Raman active vibrations in these crystals, e.g. the carboxylate symmetric stretch at 1460 cm^{-1}.[14] However, typical magnitudes of the aliphatic C-H mode scattering cross sections in these crystals are not known, nor have the linewidths been quantitatively documented. The C-H modes (2858 cm^{-1}) in the aliphatic hydrocarbon decalin have an SRS gain coefficient of 0.7 cm/GW.[13] (By comparison, the gain coefficient of the phosphate symmetric stretching mode in KDP is approximately 0.2 cm/GW.[13]) The relevance of this value to organic salts is uncertain because, among other things, the linewidths of such vibrations in crystals may be very different from those of liquid decalin. Nonetheless, it demonstrates the potential impact of these modes on SRS thresholds.

7. DAMAGE THRESHOLDS

Laser induced damage can occur through a number of mechanisms, including thermal fracture from bulk absorption,[4] local fracture at absorbing inclusions,[3] or bulk photochemistry such as photorefractive damage or color center formation. For ICF applications the primary form of damage of concern is damage at absorbing inclusions.[3]

While the favorable bulk damage properties of organic materials are sometimes cited in the literature, little systematic evidence for high damage thresholds measured under well characterized conditions has been reported. However, damage tests of LAP and its deuterated analog, LA*P, have been done at LLNL and the results are shown in Table 3.

For comparison, results obtained under the same conditions in KDP and KTP are included. It is not known whether the high damage threshold of LAP vis à vis KDP is due to a smaller number of damaging inclusions, or because the nature of the inclusions is different, or because LAP has better resistance to local fracture. Nonetheless, these results are convincing evidence that high damage thresholds can be found among organic crystals in the class we are examining. Although KTP is also capable of very high damage thresholds, scaling of this material to large sizes has not proven feasible.

Table 3.   Damage Thresholds of Nonlinear Crystals for 1 ns Pulses at 1.064 mm

Crystal	Available size (cm^3)	J_D(J/cm^2)
KDP	10x10x10	5 ± 1
LAP	1x2x3	10 ± 2
KTP	0.5x0.5x0.5	13 ± 2

8. CONCLUSIONS

Among the group of ionic organic crystals are a significant number of materials which have moderately large nonlinearities (1-2 pm/V), favorable phasematching properties for generating light in the visible or near ultraviolet, and which grow easily into large high optical quality crystals. No intrinsic constraint on the damage threshold has been discovered yet, and effects due to the nonlinear refractive index are not expected to be any worse in this class than in materials with similar linear refractive indices. Although limitations caused by two photon absorption and stimulated Raman scattering have not yet been assessed with certainty, these materials hold considerable promise as candidates for use in high power frequency conversion.

9. ACKNOWLEDGMENTS

We would like to thank R. Adair and L. Chase for providing us with the results of their n_2 measurements, and D. Milam for the damage testing results on LAP and KTP.

10. REFERENCES

1. J. Holzrichter, E. M. Campbell, J. D. Lindl, and E. Storm, "Research with High-Power Short-Wavelength Lasers," Science 229, 1045 (1985).
2. D. Eimerl, "Frequency Conversion Materials from a Device Perspective," Proc. SPIE 681, 2 (1986).
3. D. Milam, "Laser Damage in Optical Crystals," in The Laser Program Annual Report, UCRL-50021-85, (1986).
4. D. Eimerl, "High Average Power Second Harmonic Generation," IEEE J. Quant. Electron. QE-23, 575 (1987).
5. J. F. Nicaud and R. J. Tweig, "Design and Synthesis of Organic Molecular Compounds for Efficient Second Harmonic Generation," in Nonlinear Optical Properties of Organic Molecules and Crystals Vol.1, D.S. Chemla and J. Zyss, eds., Academic Press, Inc., Orlando, Florida (1987).
6. S. Kurtz, "Measurement of Nonlinear Susceptibilities," in Quantum Electronics, Vol.1A, H. Rabin and C. Tang, eds., Academic Press, New York (1975).
7. S. Velsko, "Direct Assessment of the Phasematching Properties of New Nonlinear Materials," Proc. SPIE 681, 25 (1986).
8. C. Chen, Y. X. Fan, R. C. Eckardt, and R. L. Byer, "Recent Developments in Barium Borate," Proc. SPIE 681, 12 (1986).
9. M. Szyper and P. Zuman, "Electronic Absorption of Carboxylic Acids and their Anions", Anal. Chim. Acta 85, 357 (1976).
10. S. Velsko and D. Eimerl "New SHG Materials for High Power Lasers," Proc. SPIE 622, 171 (1986).
11. D. Eimerl, S. Velsko, L. Davis, F.Wang, G.Liacono "Deuterated L-Arginine Phosphate: A New Efficient Nonlinear Crystal," in preparation.
12. R. Adair and L. Chase, private communication.
13. W. L. Smith, M.A. Henesian, and F.P. Milanovich, "Spontaneous and Stimulated Raman Scattering in KDP and Index-Matching Fluids," in The Laser Program Annual Report, UCRL-50021-85, (1986).
14. DMS Raman/IR Atlas of Organic Compounds, B. Schrader and W. Meier, eds., Verlag Chemie, GmBH (1974).

AUTHOR INDEX